Monographs on Endocrinology

Volume 3

F. G. Sulman

Hypothalamic Control of Lactation

In Collaboration with

M. Ben-David, A. Danon, S. Dikstein, Y. Givant
K. Khazen, J. Mishkinsky-Shani, I. Nir, C. P. Weller

With 58 Figures

Springer-Verlag Berlin · Heidelberg · New York 1970

Department of Applied Pharmacology
The Hebrew University
School of Pharmacy
Jerusalem, Israel

ISBN-13: 978-3-642-46235-1 e-ISBN-13: 978-3-642-46233-7
DOI: 10.1007/978-3-642-46233-7

Softcover reprint of the hardcover 1st edition 1970
Title No. 3973

Preface

The contents of this monograph "Hypothalamic Control of Lactation" are the result of studies by a nine-man team which has worked together for nine years. The topic is so fascinating that new research students are attracted to it every year, and the circle is growing steadily. This is, in fact, part of a much more extensive research field to which our Department is applying itself at present, *viz.* "Hypothalamic Pharmacology".

Our work in this sphere was made possible by a generous grant from the United States Department of Agriculture, which helped us during the first 5 years of research with Grant No. FG-IS-147, Project No. A10-AH-3, and recently with another grant for a further 5-year period, No. FG-IS-259, Project No. A10-AH-33. Our thanks are due to Dr. Joel Bitman, Hormone Physiology Leader of the USDA Agricultural Research Service at Beltsville, Md., the sponsor of this research.

As this monograph contains mainly references connected with our own work, the reader who wishes to make a survey of lactation should study the references cited in the bibliography, such as Cowie (1961, 1966), Cowie and Folley (1957, 1961), Folley (1952, 1956), Jacobsohn (1961), Lyons and Dixon (1966), Meites (1959 a, 1961, 1966), Meites and Nicoll (1966).

In this monograph the individual team members have each reported on their own special field of research. Prof. H. Z. Winnik, Head of our Psychiatric Department who has instigated, and Prof. J. Zander, Head of the Universitäts-Frauenklinik in Munich, who has encouraged our studies, deserve our special gratitude. Our work would not have been possible without the untiring devotion of Mrs. Sheila Woolfson.

Special thanks are due to Miss Yocheved Sussmann, who has been editing and proof-reading our papers for many years past and has now devoted herself to putting this monograph into proper shape. Finally, we wish to thank Dr. H. Götze and Miss E. Pfisterer of the Springer-Verlag Heidelberg, for their invaluable advice and guidance.

Jerusalem, August 1970 F. G. Sulman

Contents

Part I
Physiological Regulation of Lactation

Part II

Pharmacological Regulation of Lactation

Part III
Prolactin Assay

Part IV
Problems of Lactation

Abbreviations

The following abbreviations were used in this book:

ACTH	=	Adrenocorticotropic Hormone
AP	=	Anterior Pituitary
BIT	=	Barbitone Sleep Induction Time
BTD	=	Body Temperature Drop
CPE	=	Cataleptic Effect
CRF	=	Corticotropin (ACTH)-Releasing Factor
FRF	=	FSH-Releasing Factor
FSH	=	Follicle Stimulating Hormone
HCG	=	Human Chorionic Gonadotropin (LH)
HGH	=	Human Growth Hormone (STH)
HPL	=	Human Placental Lactogen
IU	=	International Unit
LH	=	Luteinizing Hormone
LRF	=	LH-Releasing Factor
LTH	=	Prolactin, Lactotropic Hormone, Luteotropic Hormone
ME	=	Median Eminence: The center of the portal vessel-chemotransmitters (releasing or inhibiting factors). It consists of a neural component (the infundibulum of the hypothalamus), a vascular component (the hypophysial-portal capillaries and veins) and an epithelial component (the pars tuberalis of the adenohypophysis).
MIF	=	MSH-Inhibiting Factor
MRF	=	MSH-Releasing Factor
MSH	=	Melanophore Stimulating Hormone
MTI	=	Mammotropic Index (cf. Chapter 26)
MU	=	Mouse Unit
PIF	=	Prolactin-Inhibiting Factor
PMS	=	Pregnant Mare Serum Gonadotropin (FSH)
PRF	=	Prolactin-Releasing Factor
PTS	=	Eyelid Ptosis
PTU	=	Propylthiouracil

RPL = Rat Placental Lactogen
RU = Rat Unit
SAR = Structure-Activity Relationship
SRF = STH-Releasing Factor
STH = Somatotropic Hormone, Growth Hormone
T-3 = Triiodothyronine
TRF = TSH-Releasing Factor
TSH = Thyroid Stimulating Hormone

Part I

Physiological Regulation of Lactation

Chapter 1

Mechanism of Lactation

The mechanism of lactation is poorly understood. For this reason no hormonal preparation for stimulation of lactation in man or animals has as yet been accepted. The following working hypothesis and nomenclature is proposed by us:

Stages of Lactation	*Active Hormones*
Preparation of milk gland Mammogenesis	Pituitary and later placental FSH and LH (in rodents LTH) Estradiol and progesterone
Initiation of lactation Colostrogenesis	Prolactin-inhibiting factor depressed Prolactin-releasing factor stimulated Removal of placental hormones
Secretion of milk Lactogenesis	Pituitary LTH (prolactin) Pituitary ACTH and corticosteroids
Maintenance of lactation Galactopoiesis	Pituitary STH: intrinsic lactogenic effect and augmentation of prolactin Pituitary TSH, tri-iodo-thyronine and thyroxine
Promotion of excretion Ejection: "Let-Down"	Oxytocin: contraction of mammary myoepithelium and slight stimulation of prolactin Neurohormones inhibit "let-down"

The main problem derives from the fact that, while the common hormone reactions, e. g. ovulation etc., are brought about by hyperemia of the target organ and enzymatical changes involving low-energy processes which can be induced by one hormone alone, lactation additionally requires the action of many metabolic hormones which have to mobilize casein, butter fat, lactose, calcium etc. for the excessive requirements of milk production. It is obvious that this goal cannot be achieved by one hormone alone, i. e. prolactin, without the aid of STH, ACTH, TSH, parathyroid hormone, etc. A harmonious concurrence of all these hormones is apparently only possible through appropriate stimulation by the "conductor of such a symphony"—the hypothalamus.

With this approach in mind we studied the different hormone mechanisms involved. The three main objectives of that study can be summarized as follows:

a) contribution to the basic knowledge of the mechanism of lactation by interplay of all body hormones;

b) developing exact methods for assay of all hormones involved in the different species;

c) stimulation of lactation in animals and women postpartum.

Experimental data strongly suggest that the mode of action of the ovarian hormones on mammary tissue may be direct as well as indirect. Today we know that high doses of LTH and STH can substitute for the effect of estrogen + progesterone (cf. Chapters 5 and 20).

The action of the steroids on lactation is only comprehensible if we assume the existence of an acceptor or receptor for them, both in the brain (hypothalamus and hypophysis) and in the mammary gland. As we shall show in Chapter 19, these receptors have a striking similarity with those required for the modern psychopharmaca which elicit lactation, with one difference: Lactation elicited in women by psychopharmaca—apparently induced by impairment of a prolactin-inhibiting factor (PIF) in the posterior hypothalamus—cannot be suppressed by steroids, which, as it now seems, cannot enter a receptor occupied by specific psychopharmaca. The same holds for the Chiari-Frommel syndrome and Ahumada-del Castillo syndrome. These are called forth by a disturbance of the PIF center and do not react to steroids (Chapter 28). On the other hand, puerperal lactation is apparently induced by a prolactin-releasing factor (PRF)—its area in the anterior hypothalamus is extremely sensitive to estrogen; it can be stimulated by it and it can also be suppressed by an overdose of estrogens. Thus, it seems that a double-center theory would explain all the manifold problems of hypothalamic lactation described in this monograph. Release or inhibition of additional hypophysiotropins can further contribute to lactation (DIKSTEIN and SULMAN, 1966).

A new aspect of hypothalamic lactation is its dependence upon the temporal cortex of the brain. Bilateral lesions produced by suction in the basal portion of the temporal lobes in ovariectomized estrogen-primed rabbits resulted in milk secretion (MENA and BEYER, 1968). This effect was apparent within one week after the operation. The common area destroyed by these lesions included the entorhinal cortex and the ventral part of the amygdala. Control lesions in the neocortex and olfactory bulbs did not induce this effect. Using stereotaxic approaches, we have not been able to confirm these results which—if correct—would suggest that the temporal lobes influence the hypothalamic pituitary systems.

Summarizing, it seems that the difficulty of producing lactation by prolactin alone stems from the necessity of a harmonious interplay of all hypophysiotropins. This can only be obtained by suitable stimulation (or suppression) originating in the hypothalamus. The importance of the temporal lobe of the cortex and its ability to evoke feelings of motherhood in animals and humans (cf. Solomon's judgement: I Kings: 3: 16—28) needs further study.

Peripheral hormonal factors are no less important for regulation of mammary gland growth and function. Yet, the most vital for mammary proliferation are the ovarian hormones—estrogen and progesterone. Normal cyclic changes in these two hormonal components govern the normal cyclic changes in mammary gland development associated with estrus or menstrual cycle. The same changes can be reproduced experimentally by exogenous hormonal treatment. In this respect we can distinguish between two different types of reaction: Animals such as the mouse, rat, rabbit or cat will respond to physiological amounts of estrogen by proliferation of the duct system, while for lobulo-alveolar growth progesterone is also required. Detailed studies on the optimal amounts and proportions of the two hormones have been carried out by many researchers (for details cf. D. JACOBSOHN, 1961), and are obviously different according to the species and specific strain studied. A different kind of reaction is observed in the guinea pig, goat and cow. In these species estrogen alone has been shown to cause both ductal and lobulo-alveolar development. Closer observation reveals, however, that even here treatment with estrogen alone results in anomalous gland structures (Cowie, Folley, Malpress and Richardson, 1952). For optimal mammary growth a combination of estrogen and progesterone is essential.

In still other species, such as the bitch and the ferret, estrogen alone does not even produce ductal growth.

A further point to be remembered, apart from the species differences, are the differences in the individual response of animals belonging to the same species and strain. This applies both to natural and to experimental mammary development. Lactation studies should, therefore, preferably be carried out on estrogen-primed animals. The dose given varies between 8 and 10 μg daily per rat for 10 days. We prefer the lower dosage since we found that priming with 10 μg per day may give unspecific reactions. Thus, Nicoll, Talwalker and Meites (1960) obtained lactation after non-specific stresses, such as injection of formalin, exposure to extreme cold or heat, restraint, etc., using 10 μg for daily priming. They also reported lactation following injection of meprobamate after identical priming, whereas this substance proved negative in our set-up when only 8 μg/day was used for priming (Ben-David, Dikstein and Sulman, 1965).

In short: mammary physiology is similar in all mammals, but hormone requirements for mammary development vary according to species. For experimental mammogenesis estrogen priming at subthreshold doses is of utmost importance.

Chapter 2

Hypothalamus-Pituitary Axis

a) Hypothalamic Control of Prolactin Secretion

The theory of the anterior pituitary gland being controlled by neurohumors (Fig. 1) carried through the hypophysial portal circulation, advanced mainly by Harris (1955), and the concept of neurosecretion, pioneered by Scharrer and his group (1963), have led to the present, widely held view that the median eminence of the hypothalamus and the adjacent pituitary stalk make up a specialized neurohumoral organ whose function it is to control the anterior lobe (HARRIS and DONOVAN, 1966; BROWN-GRANT and CROSS, 1966; MARTINI and GANONG, 1966; REICHLIN, 1966). Within the median

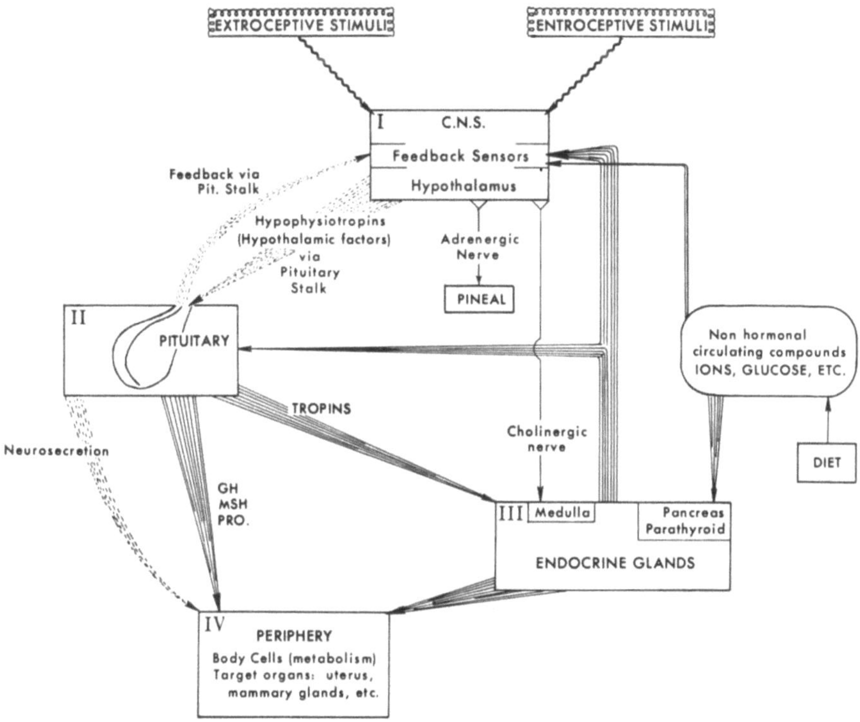

Fig. 1. Diagram of mammary stimulation by different hormones, drawn by M. BEN-DAVID (1968). Effect of I CNS (hypothalamus) and II pituitary, on III endocrine glands and IV peripheral target organs (mammary gland, etc.). Note absence of feedback from the mammary gland

eminence there are nerve endings containing neurosecretory substances which have the property of regulating the secretion of the cells of the anterior pituitary. These "hypophysiotropic hormones" or "releasers" diffuse into the capillary plexus of the median eminence and are carried by the veins of the pituitary stalk to the sinusoids of the anterior pituitary gland. This vascular system has been designated the hypophysial-portal circulation, by analogy with the hepatic-

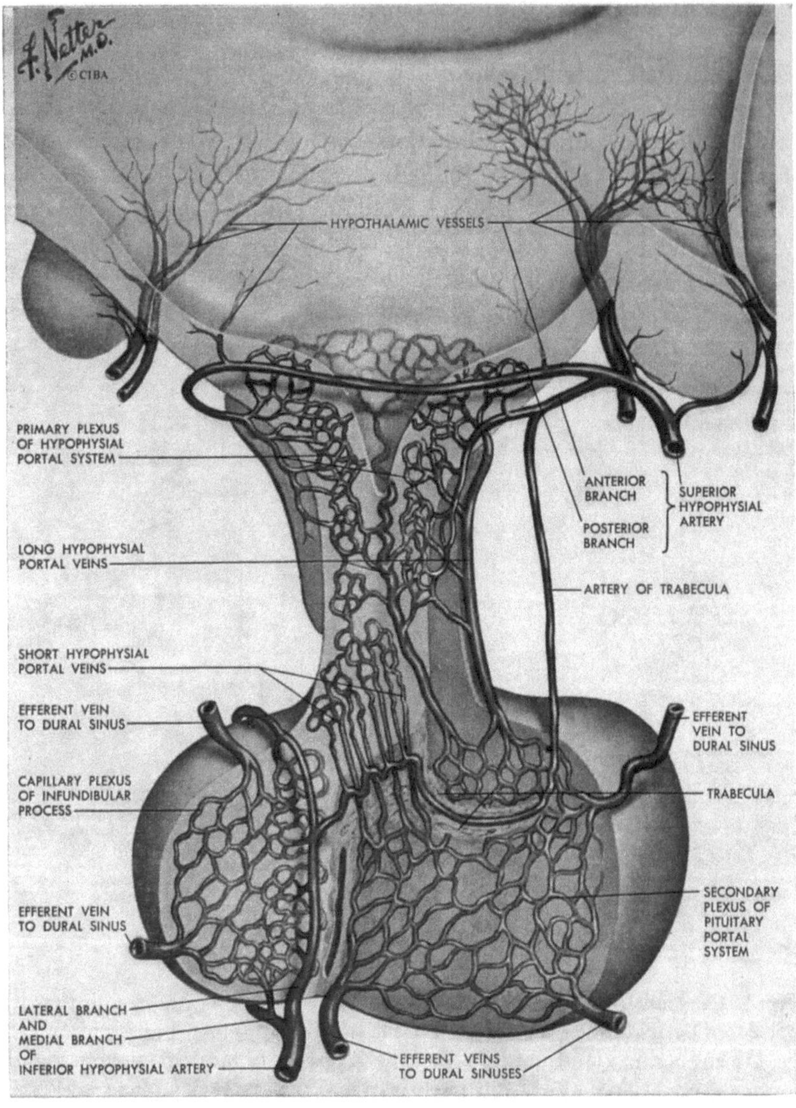

Fig. 2. Scheme of hypophysial-portal circulation. The sinusoids of the anterior lobe receive their blood supply from the hypophysial portal vessels, which arise from the capillary beds within the median eminence (above) and enter the infundibular stem (below). In the anterior lobe (r) they form a secondary plexus of the pituitary portal system, leading to the venous dural sinuses, which surround the pituitary and connect it to the general peripheral circulation. The releasing or inhibiting factors of the hypothalamus or the median eminence enter the pituitary circulation at the primary plexus of the hypophysial portal system which runs from the median eminence via the infundibular stem to the pars distalis (anterior pituitary lobe). (Copyright: The Ciba Collection of Medical Illustrations by FRANK H. NETTER, M. D.)

portal system (Fig. 2). The portal vessel chemotransmitter hypothesis has been supported by the demonstration that compounds extracted from the stalk and median eminence region of the hypothalamus possess the property of "releasing" or "inhibiting" anterior pituitary tropic hormones.

In mammals, prolactin is a unique anterior pituitary hormone because the hypothalamus exerts a tonic and chronic inhibition on its pituitary secretion by a prolactin-inhibiting factor (PIF). This fact does not exclude the possibility of the existence of a prolactin-releasing factor (PRF) as will be shown in paragraph c. In the mammalian animal the inhibitory hypothalamic regulation of prolactin secretion is the predominant one; it is, however, reversed by lactation following delivery. With regard to MSH, it seems that it, too, is subject to the regulation by a MSH-releasing factor (MRF) in the median eminence and a MSH-inhibiting factor (MIF) in the supra-optic nucleus area (TALEISNIK and TOMATIS, 1967). Research in this field will be reported in Chapter 14.

b) Prolactin-Inhibiting Factor (PIF)

A considerable body of evidence, direct and indirect, in both experimental animals and humans, has been accumulated to indicate the presence of the prolactin-inhibiting factor (PIF). Its location near the median eminence is shown in Fig. 3.

PIF activity has been found in extracts of hypothalamic fragments which inhibit release of prolactin from pituitaries cultured *in vitro* (PASTEELS, 1961 b; DANON, DIKSTEIN and SULMAN, 1963; TALWALKER, RATNER and MEITES, 1963; GALA and REECE, 1964 a; SCHALLY, STEELMAN and BOWERS, 1965). Moreover, changes in the hypothalamic PIF content during the estrus cycle of rats were recently reported by SAR and MEITES (1967). They found that pro-estrous and estrous rats had significantly less PIF in the hypothalamus than di-estrous rats. It was postulated that ovarian estrogen secreted during pro-estrus and estrus would increase prolactin release by depressing hypothalamic PIF production and by directly stimulating the pituitary. On the other hand, GALA and REECE (1964 b) reported that the concentration of PIF activity in hypothalamic extracts from rats killed in mid-lactation did not differ greatly from that in extracts from non-lactating rats. Similar findings were described by GROS-

VENOR (1965 b, c) and by GROSVENOR, McCANN and NALLAR (1965) using suckling rats for assay *in vivo*. They showed that hypothalamic, but not cerebral cortical, extracts could block the nursing-induced decline in pituitary prolactin. KUROSHIMA, ARIMURA, BOWERS and

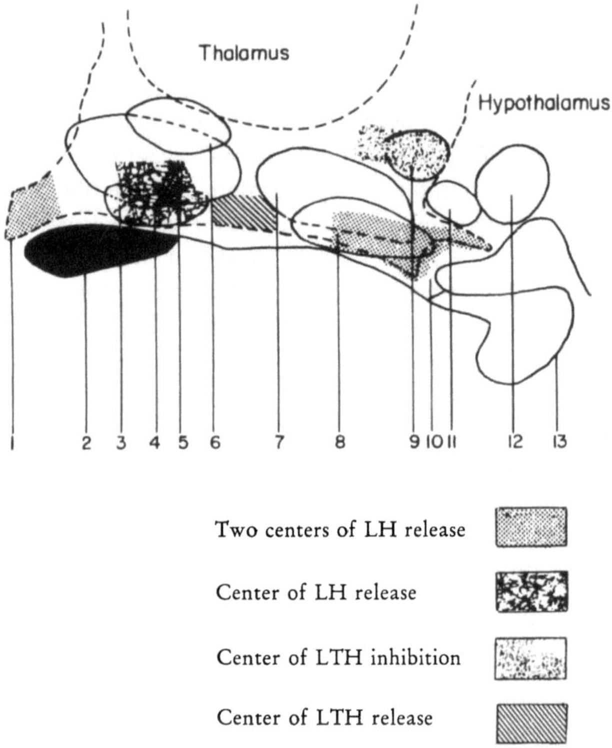

Fig. 3. Hypothalamic control of pituitary FSH, LH, LTH, indicating the mooted localization of the hypothalamic prolactin-inhibiting factor (PIF) and prolactin-releasing factor (PRF). — *1* Preoptic area; *2* Optic chiasma; *3* Suprachiasmatic nucleus; *4* Supraoptic nucleus; *5* Anterior hypothalamic nucleus; *6* Paraventricular nucleus; *7* Ventromedial nucleus; *8* Arcuate nucleus; *9* Dorso-medial nucleus; *10* Median eminence; *11* Premamillary nucleus; *12* Mamillary nucleus; *13* Hypophysis

SCHALLY (1966) demonstrated that pig hypothalamic extracts could block the depletion of pituitary prolactin which follows cervical stimulation, according to the method of HERLYN, GEELER, I. BERSWORDT-WALLRABE and R. BERSWORDT-WALLRABE (1965).

Psychopharmaca which depress the PIF center can induce prolactin release *in vivo* and initiate mammary secretion in rats, rabbits, or humans. RATNER and MEITES (1964) and MITTLER and MEITES (1967), therefore, investigated whether different lactogenic agents acted by depressing production of PIF in the hypothalamus. The effects of suckling, estradiol, reserpine, adrenaline, acetylcholine, and electrical stimulation of the uterine cervix on the content of PIF in the hypothalamus of rats were tested. After each *in vivo* treatment period, the hypothalami were removed, extracted, and tested for PIF by a 2-hour incubation with rat pituitary and subsequent assay of the medium for prolactin. Each of the agents used produced a significant decrease in hypothalamic content of PIF. DANON, DIKSTEIN and SULMAN (1963) reported that perphenazine—a chlorpromazine derivative—depressed hypothalamic inhibition of prolactin release. Both reserpine (KANEMATSU and SAWYER, 1963 a) and suckling (REECE and TURNER, 1937; GROSVENOR, 1965 a) were found to decrease pituitary prolactin content in rats, indicating release of this hormone. Some agents may elicit an increase in hypothalamic release of PIF and thereby decrease pituitary prolactin release. Thus the odor of strange males interrupts early pregnancy in mice (BRUCE and PARKES, 1963), and this phenomenon is presumably mediated through inhibition of prolactin release. If such be the case, PIF production by the hypothalamus may be increased.

The prolactin-inhibiting activity of hypothalamic extracts is not obtainable by oxytocin, vasopressin, epinephrine, acetylcholine, serotonin, histamine, substance "P", or bradykinin, suggesting that the PIF is different from any of these substances (DHARIWAL et al., 1968). The chemical nature of PIF has not been identified, although it has been observed that it passes through a semipermeable membrane, which suggests that it is a small molecule (MEITES, 1966). Prolactin-inhibiting factor activity was not found in purified extracts of the luteinizing hormone-releasing factor (LRF), indicating that these are separate factors (SCHALLY, MEITES, BOWERS and RATNER, 1964). Assays of several samples of corticotropin-releasing factor (CRF) did not show any PIF activity either (MEITES, 1966).

According to the PIF theory, all stimulating effects can only be indirectly explained by "initiation of inhibition". If the connection between the hypothalamus and the pituitary is interrupted, prolactin secretion will be invariably increased. Such an interruption has been produced experimentally by hypothalamic lesions, pituitary stalk section, pituitary transplantation, *in vitro* incubation of the pituitary together with the hypothalamus (Chapter 25) and treatment with hypothalamic depressant drugs (Chapters 13—18). Interruption of the normal connection between the hypothalamus and the pituitary may induce one of the following effects: changes in pituitary prolactin content, prevention of mammary involution or induction of mammary growth, induction of spontaneous milk secretion, maintenance of luteal function, induction of persistent pseudopregnant state with formation of deciduomata in the traumatized uterus, and, finally, induction of continuous diestrus pattern of the vaginal smear. All these phenomena indicate hypersecretion of pituitary prolactin. Therefore it is believed that accelerated synthesis and/or release of LTH is brought about by elimination of the prolactin-inhibiting factor (PIF) of the hypothalamus.

Little is yet known about the chemical nature of PIF, nor has this factor definitely been shown to exist as a separate entity in the hypothalamus. The question of whether the "luteinizing hormone-releasing factor" (LRF) and PIF are similar has been of interest ever sinve EVERETT (1956) suggested that the same hypothalamic region might control release of both LH and LTH. This is based in part on the many observations that under some conditions LH release may be inhibited when LTH secretion is high. Though purified preparations of LH-releasing factor were found to be free PIF by SCHALLY et al. (1964), activity of the latter factor was detected in preparations of the former by DHARIWAL, GROSVENOR, ANTUNES-RODRIGUES and MCCANN (1965). Different assay procedures for PIF were employed by the two groups of workers, and further purification of these factors will be necessary before this question can be resolved.

Attempts to demonstrate a direct prolactin releasing activity in intact rats *in vivo* have so far failed (RATNER, TALWALKER and MEITES, 1965). GROSVENOR (1965 a) carried out ingenious experiments in which the injection of rat hypothalamic extracts inhibited the fall in pituitary prolactin which followed 30 min nursing or stress. The specificity and reliability of this method for the measurement of PIF have, however, been questioned (MINAGUCHI and MEITES, 1967), since the failure of intact rats to respond to exogenous PIF could be attributed to the constant endogenous hypothalamic inhibition of prolactin release from the pituitary. Depression of this inhibitory hypothalamic control by a phenothiazine derivative would eliminate this difficulty and furnish a suitable set-up for the detection of exogenous prolactin-inhibiting activity.

This reasoning led DANON, DIKSTEIN and SULMAN (1970) to devise a new assay method for the *in vivo* detection of PIF: Male rats, weighing 200 ± 20 g, were injected s. c. with 10 mg/kg perphenazine, followed 30 min later by an i. p. injection of 30 mg/kg sodium pentobarbitone to induce anesthesia. One hour after the injection of perphenazine, a common carotid artery was cannulated. Then 1 ml of neutralized acid extract from 1/2 or 2 hypothalami, respectively, was infused into the artery at a rate of 0.1 ml/min. 20 min later the rats were killed by decapitation, their pituitaries were removed and homogenized in a glass homogenizer in ice-cooled 0.9% NaCl solution. The homogenate was centrifuged and the supernatant frozen for the prolactin assay. Control rats were infused with extracts of cerebral cortex prepared in the same manner.

Perphenazine injected 1 hour prior to intracarotid infusion of cerebral cortical extract significantly reduced the prolactin content of the anterior pituitary ($P < 0.0025$) as compared with intact rats infused with similar extracts of cerebral cortex. The crop gland test revealed a 50% reduction in the pituitary prolactin content. On the other hand, intracarotid infusion of hypothalamic extracts into perphenazine-treated rats elevated the pituitary hormone content as compared to that of animals infused with cerebral cortical extract. As the increase in pituitary prolactin content induced by hypothalamic extracts was dose-dependent in the range tested, it can be used for measurement of PIF levels in the hypothalamus extracts. The response produced by 1/2 hypothalamic equivalent is significant at the level of $P < 0.05$, while 2 hypothalamic equivalents evoke a highly significant increase ($P < 0.0005$).

Storage of PIF in the hypothalamus of perphenazine-treated rats has been described by DANON and SULMAN (1970). Diagram 1 (p. 12) explains the set-up. It represents the results of 3 identical experiments, the results of which were pooled and the combined data evaluated as to their significance. Analysis of these results shows that the hypothalami taken from rats 1 hour after injection of 10 mg/kg perphenazine were more effective in increasing the pituitary prolactin content of assay rats pre-treated with perphenazine and pentobarbital sodium than were hypothalami taken from vehicle-

Diagram. *1. Effect of acute treatment with perphenazine on hypothalamic PIF content in rats*

Procedure:

Stage I: Hypothalamic blockage: s. c. injection of perphenazine 10 mg/kg.

Stage II: Anesthesia: i. p. injection of pentobarbital sodium 30 mg/kg, to permit cannulation of carotid artery.

Stage III: Assay of PIF: intracarotid infusion of hypothalamic extract, or cerebral cortex extract as control.

Stage IV: Assay of pituitary prolactin: rats killed and pituitaries removed for prolactin assay.

Results expressed as average differences between mucosal reactions of two hemicrops (15 pigeons for one assay point) injected with pituitary extracts from control and experimental rats (10 rats for one assay point).

No.	Experimental set-up	Left hemicrop: control		Right hemicrop: experimental		Average difference	P
		Treatment of assay rats	Average mucosal dry weight, mg	Treatment of assay rats	Average mucosal dry weight, mg	mg	
1	Effect of hypothalamus versus cerebral cortex	Cerebral cortex extract	5.25	Hypothalamic extract	6.00	0.75 ± 0.27	< 0.01
2	Effect of hypothalamus from perphenazine rats versus hypothalamus from control rats	Hypothalamus from control rats	6.47	Hypothalamus from perphenazine-treated rats	7.43	0.96 ± 0.33	< 0.005

treated rats (Exp. 2). The difference between the two hemicrop reactions was highly significant ($p < 0.005$). A control series, in which the response of assay rats to hypothalamic extracts from intact rats was compared with the response of other rats to cerebral cortical extracts, demonstrated the sensitivity of the assay procedure (Exp. 1). Hypothalamic PIF activity resulted in elevation ($p < 0.01$) of the pituitary prolactin content of assay rats, previously depleted by perphenazine.

The increase in PIF content of the hypothalamus, while pituitary prolactin content decreases and blood-plasma prolactin rises (Ben-David and Chrambach, 1970), can best be interpreted as being due to blockage of PIF release from the hypothalamus. This results in accumulation of PIF in the hypothalamic neurons. The apparent contradiction of the present results with an earlier report showing depletion of PIF after long-term treatment (Danon et al., 1963) is easily resolved on the basis of the different treatment schedules used in the two cases. It becomes obvious that perphenazine blocks the release of PIF from the hypothalamus in the acute phase, whereas synthesis is inhibited only later, as evidenced after 10 days' treatment. Whether the impaired synthesis of PIF is directly affected by perphenazine, or is secondary to accumulation of this factor in the hypothalamic neurons, is now being studied by us. Still, the first-mentioned possibility appears more acceptable, otherwise one would expect hypothalamic PIF content not to fall beyond normal values after long-term treatment.

c) Prolactin-Releasing Factor (PRF)

The equilibrium between PIF and a tentative PRF has been a matter of controversy for a long time. As originally shown by Kragt and Meites (1965), control of prolactin release in the bird seems to differ fundamentally from the pattern observed in mammals. Pituitary autotransplants in the pigeon fail to stimulate the crop sac (Bayle and Assenmacher, 1965), and the pigeon pituitary does not release appreciable amounts of prolactin when cultured *in vitro* (Gala and Reece, 1965 a, b). In the bird, a stimulatory hypothalamic influence on prolactin secretion seems to operate which is mediated by a prolactin-stimulating factor (Kragt and Meites, 1965). PRF has been found in the hypothalamus of pigeons, chickens, quails, blackbirds, turkeys and ducks (Nicoll, 1965; Meites, 1966).

Release of prolactin by the pigeon pituitary appears to be controlled differently than in mammals. Injections of reserpine, chlorpromazine, epinephrine, acetylcholine, estrogen, progesterone, or testosterone did not, as in rats, have any effect on pituitary prolactin content or release in pigeons (MEITES and TURNER, 1947; KRAGT and MEITES, 1965); nor did cultures of pigeon pituitary show increased release of prolactin as is the case with the mammalian pituitary (NICOLL and MEITES, 1962 c). Therefore, KRAGT and MEITES (1965) studied the effect of pigeon hypothalamic extract on the release of prolactin by the pigeon pituitary *in vitro*. Prolactin release into the medium by the pituitary incubated with hypothalamic extract from parent pigeons was 2—4 times greater than that released by the pituitary incubated with cerebral cortical extract or with no extract. Slight stimulation of prolactin release was also found in the hypothalamic extract from 4—6-week-old pigeons, although this was not significant at the dose level tested. Hypothalamic extract from rats or pigeons had no effect on release of prolactin by the incubated pituitary of the other species, indicating that the hypothalamic factors regulating prolactin release in two species were not the same. These observations suggest that the pigeon hypothalamus contains a prolactin-stimulating factor in contrast to the presence of a prolactin-inhibiting factor in the mammalian hypothalamus.

Earlier studies have shown that injections of crude acid extracts of rat hypothalamus initiated lactation in rats (MEITES, TALWALKER and NICOLL, 1960 a), indicating that such extracts stimulate rather than inhibit pituitary prolactin secretion *in vivo*. The use of highly purified PIF extracts, free of other anterior pituitary regulating neurohumors, would obviate any criticism of *in vivo* methods for testing PIF activity, but such extracts are not yet available. The *in vitro* procedure has the advantage of measuring only the direct effect of PIF in hypothalamic extracts on pituitary prolactin release, since it is not complicated by the presence of target organs and is uninfluenced by other hypothalamic agents.

There is as yet no direct evidence for the existence of a PRF in mammals, though the result of several studies tend to favor it: one of these deals with the inability of suckling to influence the PIF content in the rat hypothalamus (GROSVENOR, 1965 c); another study—on the prolactin-releasing capacity of hypothalami from suckling rat mothers—carried out by MISHKINSKY, KHAZEN and

SULMAN (1968) (Chapter 23), demonstrated enhanced mammary development in rats injected with pituitaries of suckling rat mothers. DESCLIN and FLAMENT-DURAND (1969) and NICOLL (1970) also claim evidence for the existence of a hypothalamic PRF.

Another indication for the existence of a PRF was supplied by the discovery of an internal feedback mechanism for prolactin (CLEMENS and MEITES, 1968; MISHKINSKY et al., 1968), when it was shown that there exists a certain parallelism between the mechanisms of internal feedback of all tropins of the anterior pituitary. The tropins of the anterior pituitary, e. g. FSH, LH, ACTH, TSH and STH are known to exert a short feedback on the production and/or release of the endogenous tropins, by depressing their corresponding hypothalamic releasing centers. If this is the case for the above five tropins, it may well be the case for prolactin as well and may point to the existence of a PRF in addition to the PIF. PIF would then be active in normal individuals and PRF only post partum. The possibility of enhancing the lactogenic effect of a perphenazine-suppressed PIF by injecting hypothalamic extracts from suckling rat mothers—which may contain PRF—has been demonstrated by MISHKINSKY and SULMAN (unpublished results—Fig. 4).

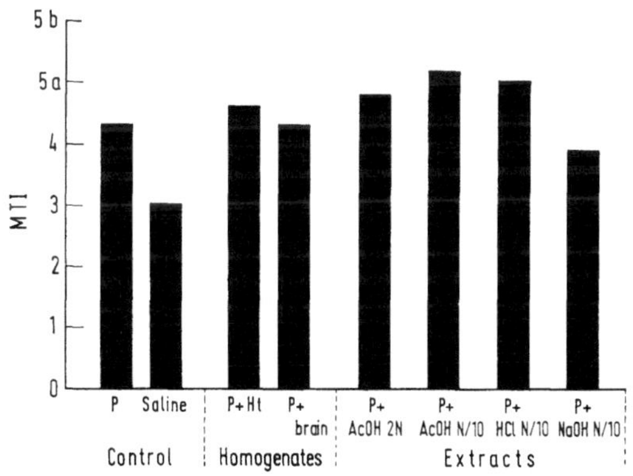

Fig. 4. Enthancing the effect of PIF suppression by perphenazine (P) followed by PRF stimulation: Average mammotropic index (MTI) of rats, after 10 days priming with estradiol (8 µg/d), followed by 5 days' injections of perphenazine (P: 5 mg/kg) and homogenates from hypothalami (Ht) taken from rat mothers 2—21 days post partum. Hypothalamus homogenate increases MTI up to 4.6; hypothalamus extracts ($N/_{10}$ acetic acid) increase MTI up to 5 a (for MTI cf. Fig. 52)

It has previously been shown that implantation of estrogen into the basal tuberal hypothalamus will evoke prolactin secretion in both rats (RAMIREZ and McCANN, 1964) and rabbits (KANEMATSU and SAWYER, 1963 b). TINDAL, KNAGGS and TURVEY (1967) later reported that estrogen implanted in the amygdala or stria terminalis will evoke prolactin release in the rabbit. These results also point to the existence of a PRF center in the hypothalamus.

Another approach to research on the releasing mechanism for prolactin arose from papers by McCANN, MACK and GALE (1959) and AVERILL (1965) in rats and by BAN, SHIMIZU and KUROTSU (1958) in rabbits, reporting cessation of lactation by specific hypothalamic lesions. Both McCANN and AVERILL postulated that bilateral destruction of hypothalamic tissues in the lateral pre-optic area and the lateral hypothalamic area, immediately posterior to the optic tracks, interferes with the production or release of prolactin from the anterior lobe of the pituitary. In 1962 TURNER and GRIFFITH advanced a tentative theory for the hormonal release of LTH, assuming that the hypothalamus contains a "lactogen-releasing factor". TURNER postulated that under normal conditions this factor builds up in the hypothalamus, producing LTH storage in the anterior pituitary, and through the suckling stimulus "this hypothalamic factor is discharged into the portal blood system to stimulate discharge of LTH from the anterior pituitary".

Similar observations were made by ANDERSON and McCANN (1955) in goats during attempts to locate the hypothalamic center of milk ejection. They confirmed ANDERSON's previous data on the relation between milk ejection and the supra-optic nuclei, and established that stimulation of these nuclei evoked milk ejection (ZAKS, 1962). Using ANDERSON's methods, POPOVICH (1958) showed that, when the supra-optic nucleus is stimulated in goats, the magnitude of milk ejection is dependent on the strength and duration of the stimulus: if the stimulus is weak—milk ejection can easily be obtained several times in the course of the same experiment. POPOVICH reasons that this inhibitory effect is not associated with the depletion of adrenaline. After destruction of one of the supra-optic nuclei by electrocoagulation, milk ejection declines on the side of the destruction, and is maintained on the intact side. After some time the reflex is partially restored on the side of the destruction (ZAKS, 1962). Thus, it seems that the failure of rats and ewes to eject milk after supra-optic lesions is due to some hormonal control of the anterior hypothalamus, since (1) oxytocin injections are not followed by milk "let-down"; (2) the histology of the mammary glands of the lesioned animals reveals a deficiency of alveolar milk constituents; and (3) lesions which do not damage the median eminence may affect a prolactin-releasing center.

EVERETT and QUINN (1966) demonstrated by electrical stimulation of the rat hypothalamus that the areas which produce ovulation indicative of luteinizing hormone release, and those which produce pseudopregnancy indicative of prolactin release, are separable. Consequently, it would appear that different hypothalamic regions are involved in the regulation of each of these tropins. QUINN and EVERETT (1967) were able to induce delayed pseudopregnancy by suitable stimulation of the area which controls prolactin release.

These studies prove that lesions in the anterior hypothalamus may interfere with the active process of lactogenesis. Since such lesions seem to inhibit prolactin release, it was of interest to determine the effect of electrical stimulation on this anterior hypothalamic area of prolactin discharge. In sexually mature, primed rats, bipolar concentric electrodes were implanted chronically in various parts of the hypothalamus. Electrical stimulation at a frequency of 25 square waves/sec, of 0.01 msec duration and at a peak current of 10 μA (10″ on—20″ off for 60 min), caused slight mammotropic development when applied to the anterior hypothalamus (Fig. 5), and none when given given to other sites of the brain (MISHKINSKY and SULMAN, unpublished results). It seems that this site of the hypothalamus is identical with that suggested by RAJKOVITZ (1962), FLERKO and SZENTAGOTHAI (1957) and AVERILL (1965), and it may be a center of prolactin discharge due to the presence of PRF. More exact experiments to localize such a center in the anterior hypothalamus are now being carried out (cf. Fig. 3).

d) Hypothalamic Feedback of Prolactin

The functions of the pituitary and hypothalamus are interrelated through two feedback mechanisms—"internal" or "short" (Fig. 6) and "external" or "peripheral" (Fig. 7). The internal feedback operates direct between the hypophysiotropins and the releasing or inhibiting factors of the hypothalamus, or *vice versa*. The peripheral feedback is indirect, being brough about via peripheral target organs which secrete their hormones into the blood and thus affect the pituitary tropins (SULMAN, 1956 b). When the hormone output from the target organ is increased, it has a negative (inhibitory) effect on the pituitary, while its decrease has the opposite, a positive (stimulatory), effect on tropin production. In certain cases different doses of a hormone may elicit a positive or a negative effect. Most peri-

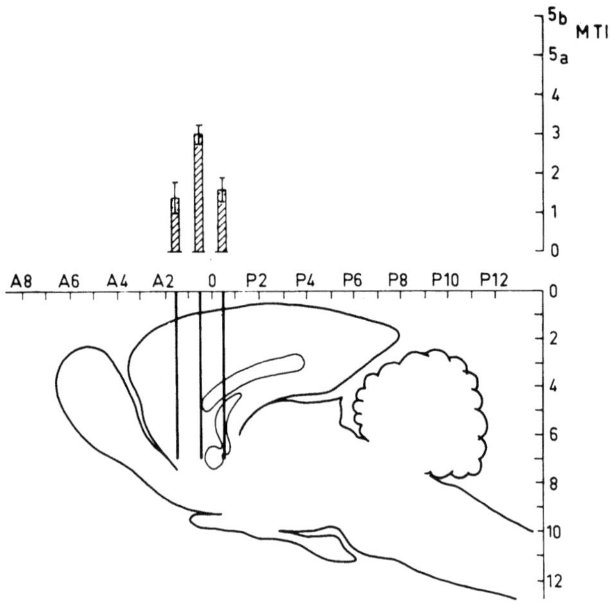

Fig. 5. Mooted center of PRF: Prolactin release following electrical stimulation of the rat brain, after estradiol priming (8 µg/d/rat) for 10 days. Prolactin release (MTI 3) is obtained in a special region of the anterior hypothalamus A-1, at a depth coordinate H-7 bilateral at L-0.5

pheral hormones, when given in small doses, are known to enhance, but in large quantities to block, the secretion of their pituitary tropins. Studies on implantation of most of the known tropins into the median eminence of the rat brain have been reported (FLERKO, 1966; MESS and MARTINI, 1968). The mechanism of a peripheral prolactin feedback has been discussud on many occasions (SULMAN, 1956 b), while CLEMENS and MEITES (1967, 1968) were the first to find that LTH, when implanted into the median eminence of female rats, creates a "feedback loop" by enhancing hypothalamic PIF secretion and ultimately causing a decrease in the pituitary prolactin level and mammary involution.

The work of TÖRÖK (1964), showing the existence of a two-directional blood flow in the hypophysial portal system from and to the hypothalamus, has exposed an anatomical route by which a pituitary tropin may pass from the pituitary to the hypothalamus (Fig. 6). This

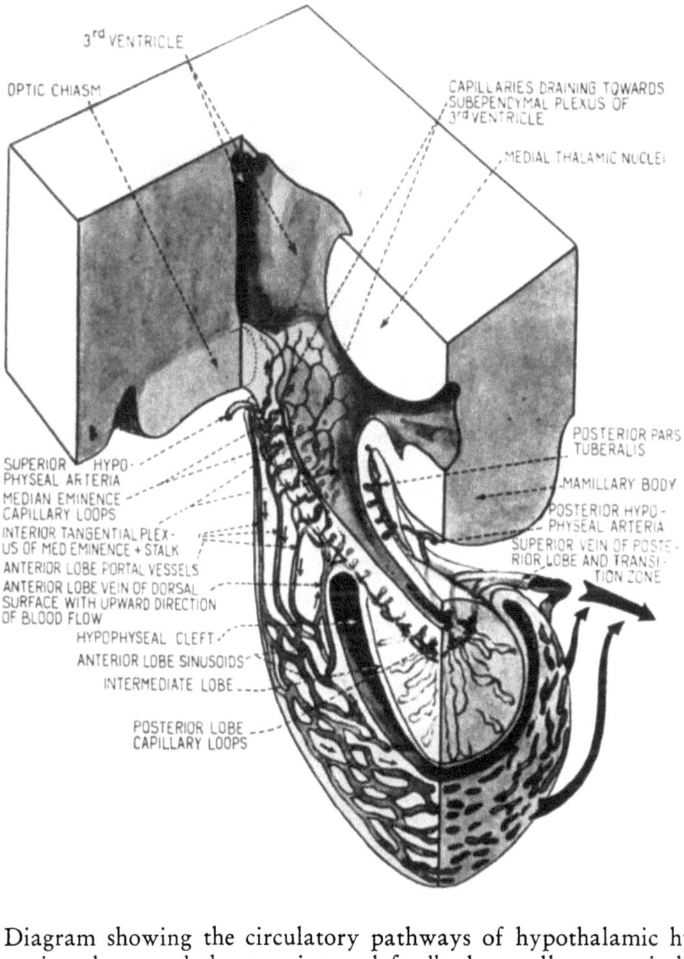

Fig. 6. Diagram showing the circulatory pathways of hypothalamic hypophysiotropin release and short or internal feedback: small arrows indicate direction of blood flow (by courtesy of Dr. TÖRÖK, 1964, Publ. House of the Hungarian Academy of Sciences, Budapest)

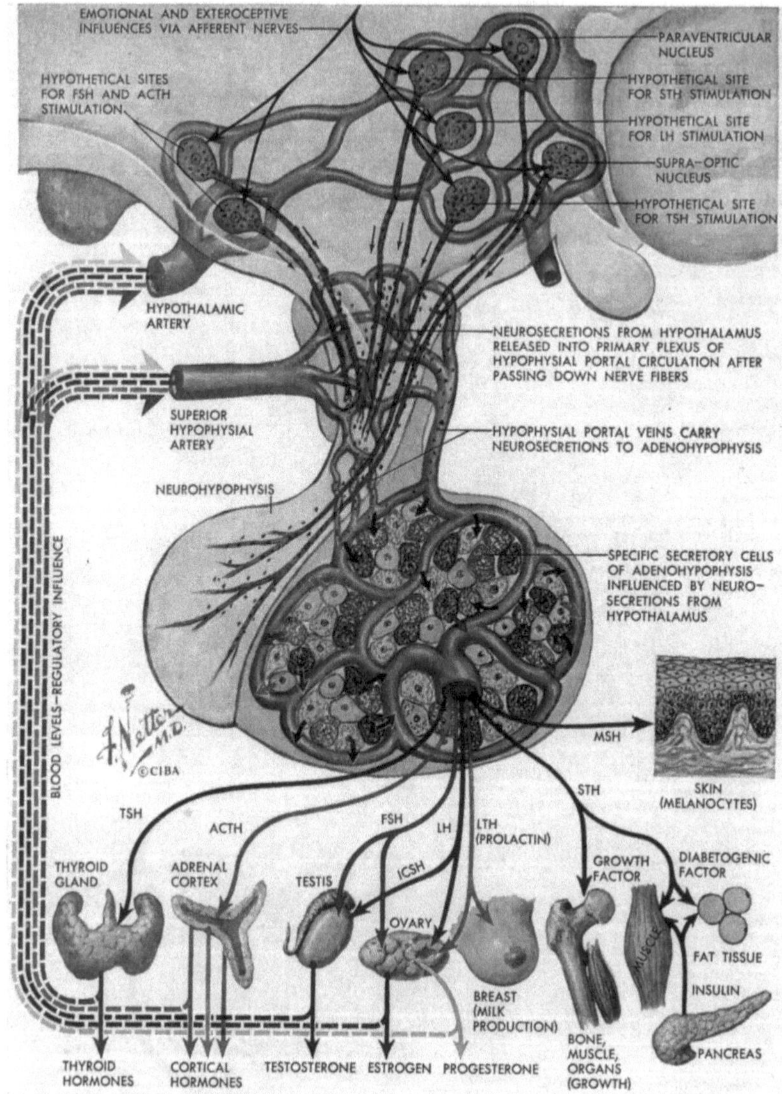

Fig. 7. Diagram showing the circulatory pathways of hypothalamus-pituitary-peripheral endocrine glands relationship: long or external feed-back. (Copyright: The Ciba Collection of Medical Illustrations by FRANK H. NETTER, M. D.)

morphological observation provides grounds for assuming that anterior pituitary hormones participate in their own feedback control (internal, short, direct or auto feedback) by acting directly on the hypothalamus. Such an internal feedback was reported for ACTH by LEGORI, MOTTA, ZANISL and MARTINI (1965), for FSH by CORBIN and STORY (1967), for LH by CORBIN and COHEN (1966) and for STH by KATZ, MOLITCH and McCANN (1967). Recently, CLEMENS and MEITES (1968) showed that implantation of ovine prolactin into the ME of mature, intact and ovariectomized rats resulted in increased hypothalamic content of PIF, reduced pituitary prolactin concentration and pituitary weight, and marked regression of the mammary glands. This effect was enhanced by simultaneous implantation of ACTH. These results suggest that the implanted prolactin inhibits pituitary prolactin synthesis and release by directly increasing hypothalamic PIF secretion. It is not known whether the small physiological amounts of prolactin reaching the hypothalamus can really exert such negative feedback, but the findings of GUILLEMIN (1967) who found pituitary tropins in hypothalamic extracts provide evidence for the existence of such an internal feedback (Fig. 7).

MISHKINSKY, NIR and SULMAN (1969) have recently shown that the internal prolactin feedback is dose-response related; this would mean that prolactin depresses its own production by exerting an inhibitory effect on the hypothalamic mechanism which in turn governs pituitary prolactin release (Fig. 8). This study was carried out on four groups of virgin female albino rats, each weighing 200 ± 10 g. After 10 days of estradiol priming, stereotaxic implantation in the median eminence of the hypothalamus was carried out. Three groups were implanted with tubes containing prolactin in doses of 1, 4 and 16 IU/rat respectively, and the fourth group, which served as control, was implanted with empty tubes. The implants were carried out in the following manner: NIH-PS-8 ovine prolactin was tapped into a micro glass tube (Rudmond's disposable pipettes) and, after all the overflow material had been cleaned from the external surface of the tube, it was immediately implanted in the ME. A follow-up showed that this technique permits complete absorption of the implanted hormone.

From Fig. 8 it is evident that the MTI values of the prolactin-treated animals are significantly decreased when compared with those of the controls: The degree of significance is $p < 0.01$ with the implantation of 4 IU, and $p < 0.005$ with 16 IU, but only $0.1 > p > 00.5$

with 1 IU of prolactin. Comparing the responses to the various doses of prolactin, a highly significant difference ($p < 0.005$) between 1 IU and 4 IU, but no significant difference between 4 and 16 IU is discernible.

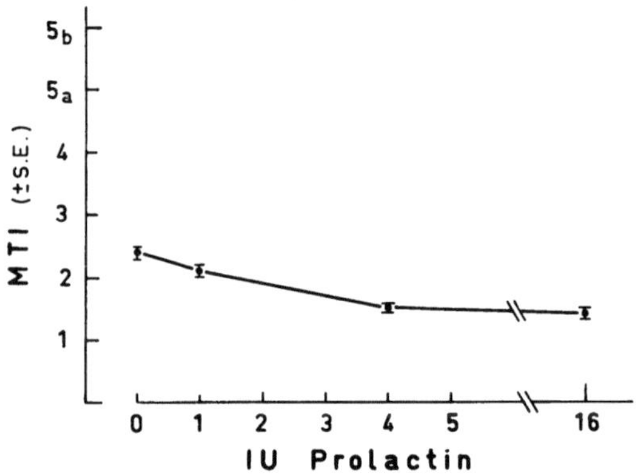

Fig. 8. Diagram of short or internal feedback after hypothalamic implantation of prolactin. Average MTI values of estradiol-primed female rats, following stereotaxic implantation of various doses of prolactin in the median eminence

For many years LTH and MSH have been considered unique among the pituitary tropins, their release being governed by a permanent inhibiting effect of the hypothalamus, in contrast to all other known tropins which were shown to be controlled by hypothalamic-releasing factors. These results have been attributed to a prolactin-inhibiting factor (PIF), and increased prolactin release has always been explained by a depression of this factor (MEITES and NICOLL, 1966; MISHKINSKY, NIR and SULMAN, 1969). Moreover, no feedback mechanism has ever been found to exist between the peripheral target organ and pituitary prolactin.

In our study, decreased MTI values were obtained following implantation in the median eminence of three different doses of prolactin. Moreover, a dose-response relationship was found to exist. These results, which show the action of prolactin to parallel that of all the other anterior-pituitary tropins, are rather surprising. The depressant effect of all the other anterior-pituitary tropins implanted in the median eminence has been attributed to the fact that they block the corresponding hypothalamic releasing factor. If in the case of prolactin there exists only an inhibiting factor (PIF) in the hypothalamus of mammals, then depression of this PIF should result in an increase in pituitary prolactin content and mammary gland stimulation. As this was not the case in our experiments, we were led to assume that, beside a prolactin-inhibiting factor (PIF), a prolactin-releasing factor (PRF) also exists in the hypothalamus. PIF and PRF may balance each other. Prolactin implanted in the median eminence may have the same depressant effect on PRF as the other pituitary tropins have on their releasing factors. When the PRF is depressed, the PIF is intensified to suppress pituitary prolactin. The existence of a prolactin-realeasing factor in the hypothalamus was previously postulated in our observations on prolactin-releasing activity of the hypothalamus in post-partum rats (MISHKINSKY, KHAZEN and SULMAN, 1968).

e) Effects of Steroids

Low doses of estradiol have been reported to stimulate prolactin secretion by direct action on the pituitary (RAMIREZ and McCANN, 1964; BEN-DAVID, DIKSTEIN and SULMAN, 1964) or indirectly by way of the hypothalamus (SAR and MEITES, 1967). Other steroids, too, have been shown to affect the PIF and pituitary prolactin content: a recent study by SAR and MEITES (1968) showed that progesterone, testosterone propionate, and cortisol acetate injections into ovariectomized rats significantly reduced the hypothalamic prolactin-inhibiting factor content, increased pituitary prolactin concentration, and promoted mammary development; testosterone propionate and cortisol acetate also induced mammary secretion. The findings on pituitary prolactin concentration are in agreement with previous studies showing that the pituitaries of rats treated with progesterone, testosterone propionate and cortisol acetate contain more prolactin than the pituitaries of control rats. The results of investigations on the hypothalamic prolactin-inhibiting factor at least partially explain how these steroids act to increase pituitary prolactin concentration and to stimulate mammary growth. Since these steroids were previously found to have no direct effect on pituitary prolactin release *in vitro*, it may be concluded that they act through the hypothalamus *in vivo*. This does not exclude the possibility that progesterone may be converted partly into estrogens *in vivo* before acting on the hypothalamus.

The above results indicate that both protein molecules and steroids can affect the hypothalamic cells responsible for the production of the factors which regulate prolactin and other tropin secretion in the pituitary. We shall show in Chapters 13—19 that drugs resembling the steroids can act the same way.

f) Interrelation between Pituitary Tropins

Just as there is an interplay of the PIF and the PRF, a reciprocal relationship seems to exist between the secretion of the gonadotropins FSH and LH, on the one hand, and prolactin, on the other. Thus, in all the experimental conditions associated with increased prolactin secretion, such as pituitary stalk section, heterotropic pituitary transplantation or treatment with hypothalamic depressant drugs, there was impairment of gonadotropin secretion. Conversely, treatment with estrogens or androgens in medium doses, known to inhibit gonadotropin secretion by a negative feedback mechanism, resulted in enhanced prolactin secretion. Similar depression of gonadotropin secretion has been observed during lactation in rats and mice, resulting in the phenomenon of "lactation-anestrus".

Voogt, Clemens and Meites (1969) implanted prolactin into the median eminence of 21-day-old female rats and studied its effects on pituitary FSH secretion. Rats killed 5 or 8 days after implantation showed a significant reduction in pituitary FSH concentration as compared with control rats implanted with cholesterol, or with control rats given implants of other hormones in amounts stated to be present in the prolactin preparation. Examination of the ovaries and uterus indicated that prolactin implantation resulted in follicular and uterine endometrial growth and increased uterine weight. These findings indicate that an implant of prolactin in the median eminence results in stimulation of FSH release in prepuberal rats.

Clemens, Minaguchi, Storey, Voogt and Meites (1969) have shown that this assumption is correct. Injections of prolactin significantly advanced the onset of puberty in immature rats by about 6 days when given at 20 days of age and by 2 days when given at 25 days of age. Injections of other anterior pituitary hormones (FSH, LH, TSH, GH) in amounts specified to be present in the NIH prolactin preparations did not advance the onset of puberty. Prolactin injections into hypophysectomized rats were also ineffective. Implants of prolactin into the median eminence hastened the onset of puberty by an average of 6.6 days, whereas implants of doses of other hormones stated to be present in the prolactin preparation had no effect on the onset of puberty. The advancement of puberty by prolactin administration is believed to be mediated through the hypothalamus and to result in decreased prolactin and increased pituitary FSH secretion.

As to the mechanism of this reciprocal relationship between gonadotropins and prolactin, it was suggested by EVERETT (1956) that the same neurogenic agent responsible for the stimulation of gonadotropin secretion may inhibit the secretion of prolactin. Studies with purified LRF, however, have shown that this hypothalamic factor is free of any prolactin-inhibiting activity (ARIMURA et al., 1967).

WELLER, DANON and SULMAN (unpublished data) found that, while acute adrenalectomy only slightly impaired perphenazine-lactation, chronic adrenalectomy resulted in a remarkable inhibition of perphenazine-lactation with a concomitant absence of corpora lutea stimulation (BEN-DAVID, 1968 b). Trying to elucidate the mechanism of this phenomenon, we explanted pituitaries and hypothalami from perphenazine-treated rats after chronic adrenalectomy and ovariectomy (perphenazine treatment started 10 days after operation). These pituitaries released considerably less prolactin than pituitaries with hypothalami from perphenazine-treated rats after acute adrenalectomy-ovariectomy (perphenazine treatment started on the day following operation). We would explain these observations on the basis of the assumption that, after an established over-shoot of one tropin resulting from extirpation of its target organ, the pituitary is unable to increase its prolactin secretion in spite of a proper stimulans. The mechanism of such a phenomenon could also be mediated through the hypothalamus. Measurement of hypothalamic releasing or inhibitory factors along with the pituitary tropins in these experiments could provide an answer.

Summarizing, one would assume that reactions ascribed to the PIF or to the PRF could also result from the interplay between different hypophysiotropins which may mimic prolactin release or suppression.

g) Suckling

Stimulation of the nipples results in rapid release of prolactin, corticotropin, and oxytocin (MEITES, 1959 a; COWIE and FOLLEY,

1961). This mechanism may also release growth hormone as has been shown by GROSVENOR, KRULICH and McCANN (1968). Earlier investigations had already shown that appetite and water intake, too, are increased by the suckling stimulus (CROSS, 1961), suggesting a possible release of GH and TSH.

It is of interest that the suckling stimulus can prolong lactation considerably beyond the usual post-partum interval if suckling and milk removal are permitted to continue. The litters of rats and mice are normally weaned at the end of about 3 weeks, but if there is regular replacement of the young with fresh litters from other mother rats or mice, lactation may be prolonged for as long as 9 months with relatively little decrease in milk yield (BRUCE, 1958; NICOLL and MEITES, 1959). Continued application of the suckling stimulus has been observed to prolong lactation considerably also in other species, including the human. Regular applications of the suckling stimulus without removal of milk, i. e. by ligating the galactophores of mice or rats while permitting litters to suckle, retards mammary involution and maintains slight secretory activity (SELYE, COLLIP and THOMSON, 1934). In the absence of both the suckling stimulus and removal of milk in post-partum rats, injections of prolactin, adrenal glucocorticoids, or oxytocin, either individually or in combination, may also retard mammary involution and sustain secretory activity far beyond the usual post-partum period (MEITES and NICOLL, 1959). Some maintenance of mammary lobulo-alveolar tissue and secretory activity was observed in post-partum rats for as long as 75 days after litter removal, by daily injections of prolactin, cortisol acetate and oxytocin. Even in hypophysectomized post-partum rats after litter removal, administration of these three hormones was effective in retarding mammary involution (MEITES and HOPKINS, 1961).

The suckling or milking stimulus results in a reflex release of oxytocin from the neurohypophysis into the circulation, which then acts to contract the myoepithelial elements surrounding the alveoli and finer ducts (CROSS, 1961). This reflex can be conditioned, and oxytocin can be released in response to psychic stimuli (Fig. 9). Electrical stimulation of the para-ventricular nucleus (ANDERSON and McCANN, 1955) or of the nerve tracts to the neurohypophysis (CROSS and HARRIS, 1952), results in milk ejection, suggesting that these sites are involved in oxytocin release. Epinephrine has been shown to inhibit the local action of oxytocin on the mammary gland (BRAUDE and

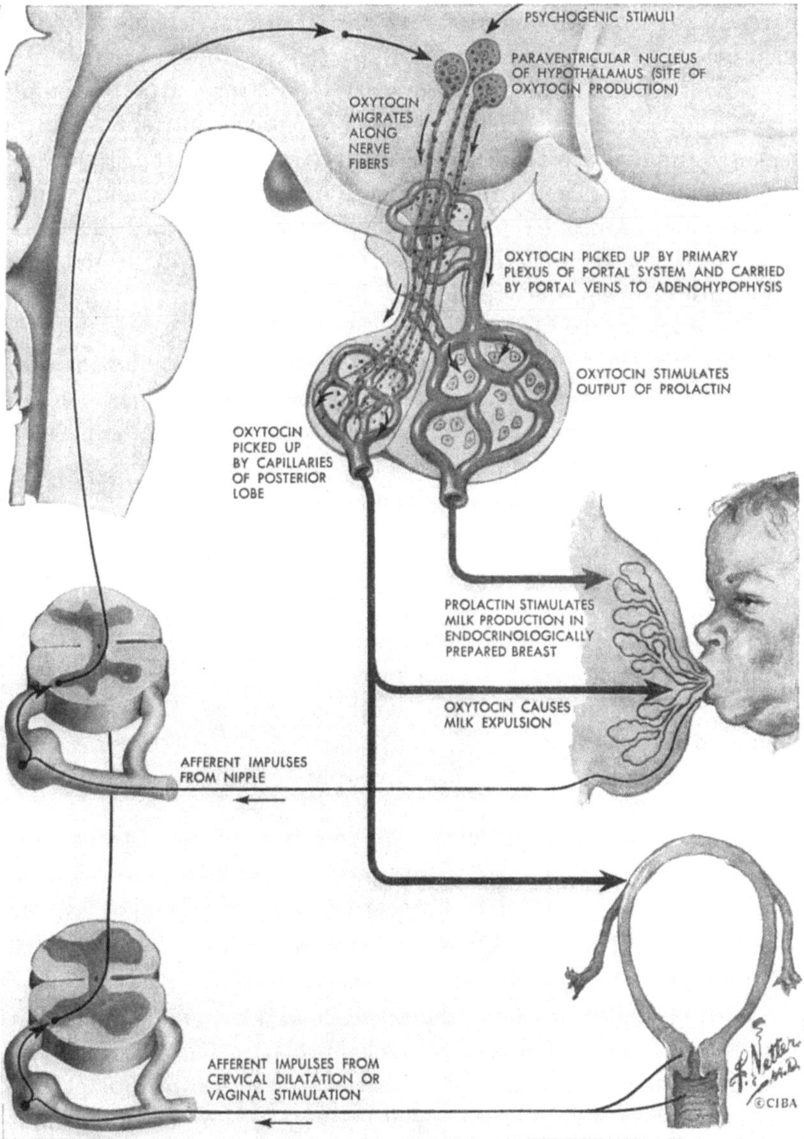

Fig. 9. Pathways of oxytocin release induced by suckling. (Copyright: The Ciba Collection of Medical Illustrations by FRANK H. NETTER, M. D.)

MITCHELL, 1952; CROSS, 1953), but emotional stress (cf. Fig. 13) apparently prevents milk ejection by a partial or complete block of oxytocin release from the posterior pituitary (CROSS, 1955).

Recent studies in Meites' laboratory (1966) suggest that serotonin may be involved in the reflex release of oxytocin in response to the suckling stimulus. Serotonin, its precursors (tryptophan and 5-hydroxytryptophan), and a monoamine oxidase inhibitor (iproniazid) significantly inhibited milk ejection; the inhibitory effects of these substances were overcome by injections of oxytocin or by a specific serotonin antagonist (2-bromo-D-lysergic acid diethylamide).

GROSVENOR (1965 b, c) reported that suckling for 30 min failed to alter the prolactin-inhibiting activity of rat hypothalamus, as compared to the nonsuckled controls. As the hypothalamus contains numerous pharmacologically active substances (e. g. serotonin, histamine, catecholamines, and acetylcholine) in large quantities, the possibility must be considered that these agents may influence the *in vivo* release of prolactin.

h) Conclusions

The fact that the hypophysiotropins are under the influence of hypothalamic releasers has been proved with regard to FSH, LH, ACTH, STH and TSH. Their synthesis, storage and secretion is regulated by releasing factors carried through the portal system into the anterior pituitary. All these hypophysiotropins are "balanced" tropins, i. e. they are also regulated by a feedback of the peripheral hormones created by them. MSH and LTH are unbalanced hormones, i. e. they do not create secondary hormones which may exert a feedback. It has been proved that MSH secretion is regulated both by a hypothalamic releasing factor and an inhibiting factor (cf. Chapter 14 — i).

With regard to prolactin, the existence of a hypothalamic inhibiting factor (PIF) is well established by prolactin release after hypothalamic lesions, pituitary stalk section, pituitary transplantation, treatment with hypothalamic depressants and by hypothalamus-pituitary co-culture.

The existence of a hypothalamic prolactin-releasing factor (PRF) is postulated by us on the basis of the following body of evidence:

a) inability of suckling to influence the PIF content of the rat hypothalamus;

b) steady increase of prolactin-releasing activity in the hypothalamus of suckling rat mothers during the 21 days of their nursing activity;

c) direct demonstration of such an increase of PRF activity in organ culture;

d) increase in the lactogenic effect of a hypothalamus depressant (perphenazine) by the addition of extracts from the hypothalamus of nursing rat mothers;

e) existence of an "internal" feedback mechanism for prolactin, which is subject to a dose-response relationship;

f) cessation of lactation after bilateral destruction of hypothalamic tissues in the lateral pre-optic area;

g) stimulation of lactation and deciduoma formation by electrical stimulation of the same area of the anterior hypothalamus;

h) stimulation of lactation by estrogen implanted into the amygdala or stria terminalis;

i) existence of a PRF in birds;

j) clinical observation that lactation due to the PRF can be abolished by steroids, whereas galactorrhea due to PIF suppression does not react to steroids and barely to clomiphene.

These findings do not, admittedly, constitute a direct proof, since oxytocin stimulation as well as a change in the balance of other auxiliary hypophysiotropins could also account for most of the observations cited. Obviously, unequivocal evidence for the existence of a PRF must depend on its chemical isolation. NICOLL has recently presented new evidence for the existence of a PRF in mammals (1970).

Chapter 3

Prolactin

a) Introduction

Interest in the physiology of pituitary prolactin greatly increased in recent years, as indicated by the large number of investigations on this subject since 1960. Prolactin is distinguished from other pituitary tropins by its inhibitory regulation originating in the central nervous system, i.e. the hypothalamus, by lack of a reciprocal peripheral feedback mechanism, by its diverse effects among the vertebrates, including growth-promoting actions in several species, and by its close association with growth hormone.

b) Storage and Release

Prolactin is probably present in the hypophysis of all vertebrates (NICOLL, 1968), but it is more plentiful in some species and at certain stages of the life cycle than at others. Ox and sheep glands usually yield 30 to 40 IU per gram of fresh tissue, while pork glands contain much less. Measurements of the prolactin content of human pituitaries indicate that the hormone accounts for about 2 percent of the acetone-dried residue of the gland. More prolactin is contained in the adeno-hypophyses of goats, sheep and cows, and less in the glands of cats and rabbits. In most species the prolactin content of the pituitary increases steadily during pregnancy, with an increase sometimes occurring soon after parturition. This is also reflected in the MTI (Fig. 10). When nursing is permitted, the prolactin content remains at this higher level for a longer period than when nursing is prevented.

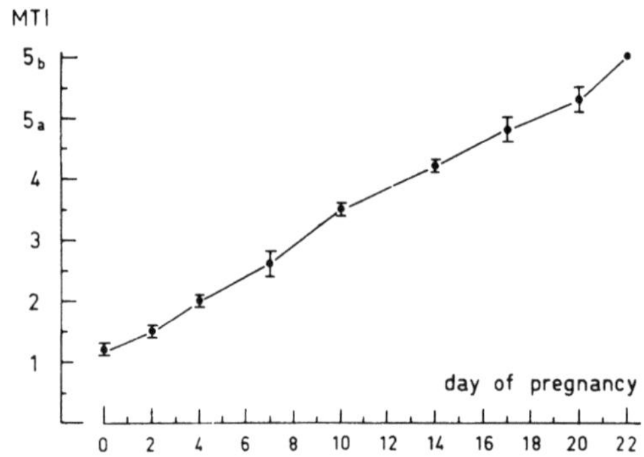

Fig. 10. Increase in mammary development during pregnancy in rats: Natural mammotropic index (cf. Fig. 52)

c) Properties

Prolactin has been prepared in a highly purified form from ox and sheep pituitaries. Isolation procedures have been described by LI (1961) in a review on the biochemistry of prolactin. The physical properties of sheep (198 aminoacids) and ox (205 aminoacids) prolactin are remarkably similar. The hormone from both species has a molecular weight of 24,000 to 26,000, an isoelectric point at pH 5.73, and a sedimentation constant of 2.19. DIXON and LI (1964) showed that the hormone molecule consists of a single peptide chain with one free amino acid group, but no free carboxy end-group. They suggested that the peptide chain has an intra-chain disulfide bridge forming a ring similar to that present in vasopressin and oxytocin. The results of their structural studies are summarized in the following schematic representation of prolactin:

H_2N-Thr-Pro-Val-Thr-Pro-$(CyS)_4$-(CyS)-Tyr-Leu-Asp(NH_2)-CyS-COOH .
1 2 3 4 5 202 203 204 205

In order to investigate aggregation and charge differences between prolactin components, a combination of gel filtration, ion-exchange chromatography, and zone electrophoresis was used by SLUYSER and LI (1964). As a result of these investigations, a method was developed for the isolation of monomer prolactin in highly purified form. The experiments presented in their paper show that the various peaks or bands which appear when prolactin is submitted to either ion exchange chromatography or zone electrophoresis are due to the presence of molecular aggregates in the preparations. The degree of aggregation differed from batch to batch. Some prolactin preparations contained about 20% monomer whereas others had a much lower content of unassociated material. In view of the reported finding that some aggregation of prolactin can occur when solutions are submitted to lyophilization, pervaporation, sucrose concentrates, or extremes of pH, it seems likely that the differences in degree of aggregation between various batches of lactogenic hormone arise at least partly during the preparation of the samples. In order to prevent aggregation, the hormone solution was concentrated by chromatography on small columns of DEAE-cellulose. A separation was achieved between fractions representing different states of polymerization of prolactin molecules. One of the fractions contained monomer prolactin, as revealed by exclusion chromatography and sedimentation equilibrium experiments. The molecular weight of monomer prolactin was estimated to be 23,000.

Prolactin is not soluble in any fat solvent. It is, however, soluble in water, except in the isoelectric region pH 5—6, and in strongly acid solution. It may be salted out with sodium or ammonium sulfate or sodium chloride, the required amount depending on the pH. It is also precipitated by basic salts of heavy metals at or near pH 7. It is rapdily inactivated by trypsin and pepsin. Salt-free solutions at pH 8 withstand boiling for one hour with only negligible loss; under other conditions, prolactin may be rapidly destroyed. Frozen preparations are not stable, once they have been thawed.

Despite the great similarities of bovine and ovine prolactin, certain differences have been observed. It seems that, proportionate to weight, ovine prolactin is more active than bovine prolactin. There is less tyrosine in ovine prolactin, and distribution of the two hormones in countercurrent analysis is different. However, BRYANT and GREENWOOD (1968) showed by radioimmunoassay of blood that ovine prolactin antigen can substitute for bovine prolactin. Human prolactin has yet to be characterized, but the fact that methods effective for isolating ox and sheep prolactin have completely failed when applied to human pituitaries suggests that primate prolactin may have peculiarities of structure which make it very similar to human growth hormone.

d) Prolactin in Body Fluids

Prolactin-like activity in urine has been demonstrated both by pigeon crop sac assay and by observation of the effect of urine on the hormonally-primed mammary gland of the immature hypophysectomized male rat. SEGALOFF and STEELMAN (1959) found daily up to 175 IU of prolactin in urine from menopausal women. In normal women the excretion was somewhat higher in the luteal phase of the cycle as compared with the follicular stage. Prolactin excretion also increased during the course of pregnancy. Prolactin-like activity of serum has been detected with both pigeon and mouse assays. The blood of lactating women has been found to contain prolactin activity (SIMKIN and GOODART, 1960). Somewhat smaller amounts of hormone were found in blood during the luteal phase of women with normal menstrual cycles. No prolactin activity was evident in the blood of women in the follicular phase of the menstrual cycle or in normal young men. These observations still require confirmation (cf. Chapter 26 — d).

e) Prolactin in Pituitary Tumors

Persistent lactation, with or without previous pregnancy, is encountered with considerable frequency in endocrine practice. The occurrence of this complaint in acromegalic women is well known. When lactorrhea and amenorrhea follow pregnancy (Chiari-Frommel syndrome), a functional disturbance in prolactin secretion may be present. General awareness that persistent lactation may also be a sign of pituitary tumor in the non-acromegalic patient followed the description in 1954 of fifteen such cases from Albright's Clinic (Forbes-Albright syndrome—cf. Chapter 28). The problem has been studied in rats bearing pituitary "mammotropic" tumors by MIZUNO, TALWALKER and MEITES (1964).

f) Effect of Prolactin on Mammary Growth

Present evidence indicates that mainly one anterior pituitary hormone—prolactin, and perhaps also STH, are involved in the development of the mammary gland. The most convincing and impressive work was done by LYONS (1942), who injected small amounts of purified prolactin (according to standards of 1942) into the mammary ducts of spayed rabbits and obtained not only clear-cut mammary development and lactational response, but also histological evidence of cellular hyperplasia of the alveolar tissue. This "lactational" growth was characterized by the presence of numerous mitoses in the alveoli and also by the fact that the number of cells per alveolus was on the average greater in the prolactin-injected sectors than in control sectors of the same gland.

COWIE and FOLLEY (1947) found that unfractionated ox anterior pituitary extracts will evoke mammary alevolar growth in the immature gonadectomized-adrenalectomized rat which presumably has no source of steroid-hormones.

LYONS (1951) elucidated experimentally the hormonal factors needed to evoke growth of the mammary lobulo-alveolar system in the hypophysectomized rat to a degree equal to that attained by the end of pregnancy. In the course of this work it was found that considerable, though not complete, growth of the mammary lobulo-alveolar system in hypophysectomized-ovariectomized rats could be evoked by suitable treatment with estrogen, progesterone and purified prolactin, the steroids themselves being ineffective in the absence of prolactin. The mammary lobulo-alveolar growth generated by this triad of hormones was not, however, as extensive as that obtained by substituting for the purified prolactin a cruder preparation known to contain ACTH and STH. Later, LYONS, LI and JOHNSON (1952) reported that the mammary lobulo-alveolar growth-promoting action of the hormone triad—estrogen + progesterone + purified prolactin—in hypophysectomized-ovariectomized rats was enhanced by addition of purified STH. Omitting prolactin from this hormonal tetrad resulted in loss of mammary alveolar growth-promoting activity, though duct growth, as evidenced by the presence of club-shaped end buds, continued.

A most interesting development during the past few years is the growing evidence that prolactin and STH in combination can induce full mammary lobulo-alveolar growth in the apparent absence of ovarian hormones in the rat. Frequent injections of relatively large doses of prolactin and STH were found to induce mammary lobulo-alveolar growth in male and female rats that had undergone gonadectomy and adrenalectomy (MEITES, 1965) or ovariectomy, adrenalectomy, and hypophysectomy (TALWALKER and MEITES, 1961). MEITES and KRAGT (1964) showed that hypophysectomized rats, with a pituitary implanted adjacent to one of the mammary glands, showed localized lobulo-alveolar mammary growth in the vicinity of the transplant.

Thus, it becomes evident that prolactin and STH together can induce full lobulo-alveolar mammary growth in the rat, which suggests that ovarian hormones are of secondary importance. The ovarian hormones are believed to promote mammary growth under natural conditions by stimulating secretion of prolactin and possibly STH by the adenohypophysis, by synergizing with these adenohypophysial hormones, and by sensitizing the mammary gland to these hormones.

g) Levels of Pituitary Prolactin in Various Stage of Lactation

SIMKIN and ARCE (1963) found no prolactin activity in the blood of children, normal young men, or normal young women in the first half of the menstrual cycle. However, prolactin activity was present in the blood of normal women during the second half of the cycle. This evaluation of the role of prolactin in the normal human menstrual cycle suggests that the time of its peak concentration in blood differs from the peaks of the other two known pituitary gonadotropins—follicle stimulating hormone (FSH) and luteinizing hormone (LH)—which fall before the midcycle (the time of ovulation). On the other hand, ample evidence has been accumulated that FSH and LH are the gonadotropins responsible for the entire process of human ovulation, icnluding the post-ovulatory development of the corpus luteum. This has been corroborated by studies on the effect of the contraceptive agents which prevent ovulation by inhibiting pituitary release of FSH and LH (cf. Chapter 28). The role of prolactin in this set-up has not yet been studied. The presence of prolactin activity during the latter part of the cycle would, however, suggest that prolactin acts as an antagonist to LH. This may be deduced from the fact that the estrus cycle of rodents is inhibited by prolactin or by a prolactin releaser (Fig. 11). Yet, it may also be due to the luteotropic effect of prolactin.

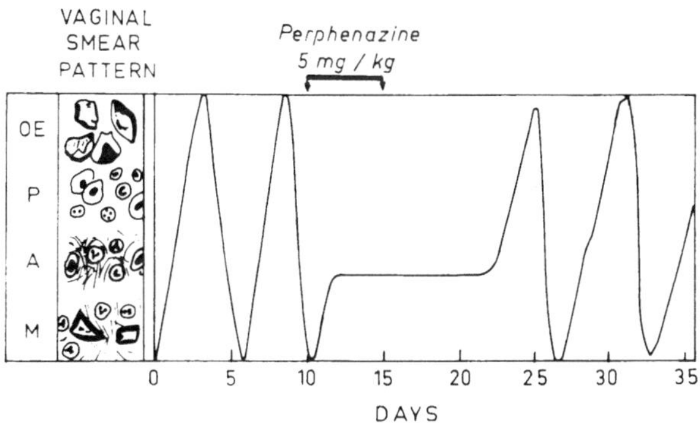

Fig. 11. Inhibition of rat estrus cycle by prolactin: Prolactin is released by perphenazine which suppresses the hypothalamic prolactin-inhibiting factor (PIF). The vaginal smear shows anestrus (A) during 14 days following injection. Thereafter, the normal cycle returns with proestrus (P), estrus (OE) and metestrus (M)

Later, in 1968, Minaguchi, Clemens and Meites showed that pituitary content and concentration of prolactin are much lower in sexually immature than in mature rats. The levels of pituitary prolactin remain low before, and at the beginning of, puberty despite increases in body and anterior pituitary weights during this period. A marked increase in pituitary prolactin content and concentration occurs soon after onset of puberty in rats (vaginal opening). These changes in pituitary prolactin levels of rats occur before and after the onset of puberty and are strikingly different from those observed for the pituitary gonadotropins. Thus both FSH and LH reach their highest levels in the pituitary prior to puberty, and show a marked decrease shortly before and after the onset of puberty. Pituitary prolactin levels showing the opposite trend provide another example of the divergence between prolactin and gonadotropin and their functions during different physiological states. The large increase in pituitary prolactin after puberty is, therefore, believed to be due mainly to increased estrogen secretion by the ovaries.

Studies of pituitary prolactin content in the rat have shown that it is low during pregnancy but increases markedly after parturition. GROSVENOR and TURNER (1958, 1960) determined the prolactin content of the pituitary in the rat at different stages of pregnancy and lactation: Prolactin content of the pituitary is low during the first part of pregnancy, later on it exceeds that of the controls and on the 16th day it reaches its peak. Just before parturition it falls to a very low level.

Pituitary histology shows similar trends. It is well known that lactogenic hormone is secreted by acidophile cells of the anterior lobe of the pituitary gland. The number and secretory activity of acidophile cells in the rat normally increases in late pregnancy and following parturition; it gradually declines as lactation advances. It is believed that the initial increase in pituitary lactogenic hormone is due to an estrogen surge which reaches a peak in blood and urine near parturition but, as the estrogen level subsides after parturition, maintenance of lactogenic hormone secretion becomes dependent upon the regular application of suckling or nursing stimuli. This conforms with the investigations of REECE and TURNER (1937) who showed that nervous stimuli resulting from milking or nursing are transmitted to the anterior lobe, causing a release of lactogenic hormone into the blood stream. The prenursing level of lactogenic hormone increased from the 2nd to the 6th day postpartum, then declined towards the end of lactation. The fall in hormone concentration as a result of 30 minutes' nursing stimuli was steepest on days 6 and 16 (95% and 90% respectively), while on the 21st day—when milk production by the rat is at its most intensive—the prolactin content of the pituitary was equal to the control level and was not affected by the suckling stimulus. The loss in prolactin content as a result of the suckling stimulus could be proved by direct determination with starch-gel electrophoresis of the rat pituitary (KURCZ, 1967; BARANYAI, NAGY and KURCZ, 1967).

h) Theories on the Mechanism of Onset of Lactation

TURNER (1939, 1952, 1956), FOLLEY (1956) and MEITES (1956, 1959 a, b) have brought forward different theories to explain the onset of lactation. However, all investigators have stressed the importance of estrogen-progesterone combinations in inhibiting milk secretion during pregnancy, an effect which this combination appears to exert by blocking the release of lactogenic hormone from the pituitary and by rendering the alveolar cells unresponsive to the lactogenic complex.

At the end of pregnancy, the fall in the rates of progesterone and estrogen removes this inhibition which is replaced by the positive "lactogenic" effect of estrogen acting unopposed (FOLLEY, 1956); alternatively, the decline in both estrogen and progesterone leaves the mammary gland receptive to prolactin stimulation and, at the same time, estrogen is still present in sufficient quantity to increase the secretion of prolactin from the pituitary and initiate lactation (MEITES, 1959 a, b).

i) Conclusion

Prolactin is the specific hypophysiotropin which promotes lactation in the pregnant mammal. Its effect seems to be dependent on adequate priming by ovarian steroids which can, however, be dispensed with by applying very high doses of prolactin fortified by STH. This combination can induce full lobulo-alveolar growth in the absence of ovarian hormones. Whereas in rodents and ferrets prolactin is luteotropic, it is an antagonist of LH in most other mammals, including the human. Its antagonism to FSH is not absolute.

During pregnancy lactation is suppressed by ovarian and placental steroids; prolactin release begins with the expulsion of the placenta. At that stage, its activity is enhanced by corticosteroids. In the cow, prolactin plays a lesser role which will be described in Chapter 27.

In male rodents prolactin stimulates the development of the prostate, seminal vesicles and preputial glands. Its effect on the mammary gland is normally inhibited by the PIF which is only overcome under pathological conditions.

Chapter 4

Growth Hormone

Early experiments demonstrated that increases in milk production could be obtained by giving injections of extracts of the anterior pituitary to lactating cows. The best responses were obtained when cows were treated during declining lactation. Experiments by FOLLEY and YOUNG (1940) demonstrated that this response was largely due to the growth hormone present in the extracts, and they showed that preparations of highly-purified bovine growth hormone gave similar increases in yield. This yield-boosting effect of growth hormone has since been confirmed by several investigators. The experiments of BRUMBY and HANCOCK (1955), who studied the effect of daily injections of purified bovine growth hormone in identical twin cows for a 12-week period over the peak phase of lactation, and a 4-week period over the end phase of lactation, showed a very substantial increase in milk yield. We have not observed this effect in rats which received perphenazine as prolactin releaser with bovine growth hormone added for milk yield boosting (Fig. 12).

Both ACTH and STH have a prolactin-sparing effect on the restoration of lactation in hypophysectomized animals (SHAW, 1955; COWIE, 1957). Adrenocortical hormones and STH may replace prolactin completely in mice (NANDI, 1958 a, b, 1959) and are at least as

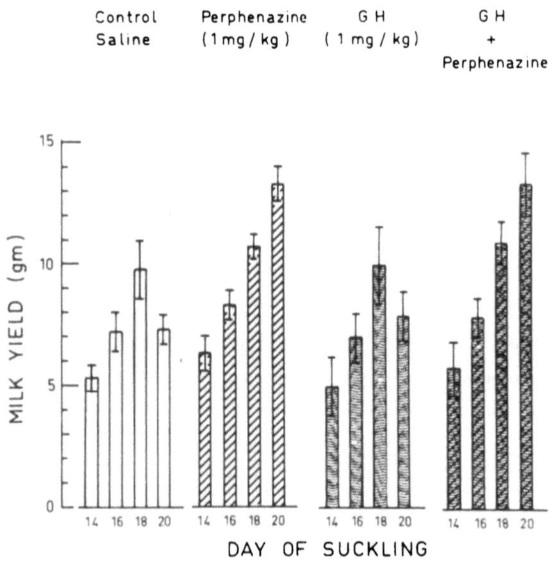

Fig. 12. Effect of growth hormone (GH) on milk yield. Perphenazine increases milk yield by prolactin release. Growth hormone does not improve milk yield

important as prolactin for maintaining lactation in ruminants. TSH and STH both appear to be necessary for maximum milk production, and they have pronounced stimulatory effects in cattle and goats (MEITES, 1959 a; MEITES, HOPKINS and TALWALKER, 1963). It has also been reported that STH can increase milk yields in rats (GROSVENOR and TURNER, 1959 a; MOON, 1965), but other workers have failed to observe this effect (MEITES, 1957 b, 1959 a; COWIE and FOLLEY, 1961).

These observations raise the question of interplay between STH and LTH in relation to the physiological phenomenon of lactation. The fact that bovine LTH injections do not raise milk yield in cows (SULMAN and TWERSKY, 1948) while STH injections do—only means that in normal conditions of lactation the amount of LTH in the blood is sufficient whereas that of STH is not. This has been proved by SCHAMS and KARG (1969) who could not detect any significant prolactin levels in the blood of cows before or during lactation. Their method of radioimmunoassay revealed, however, a significant peak one day before parturition when estrogen levels are high in the cow.

The apparent difficulty in separating prolactin from growth hormone in chemical fractionation of human pituitaries (WILHELMI, 1961; LI, 1961, 1962), and the demonstration in animals of overlapping biological activity of these two human hormones (CHADWICK, FOLLEY and GEMZELL, 1961; FERGUSON and WALLACE, 1961; LYONS, LI and JOHNSON, 1961; BARRETT, FRIESEN and ASTWOOD, 1962; BUTT, 1962) prompted an initial study by a team headed by BECK et al. (1964 a, b) on the metabolic effect in man of ovine prolactin. BERGENSTAL and LIPSETT (1958) had reported nitrogen retention in four hypophysectomized women receiving ovine prolactin, and MCCALISTER and WELBOURN (1962) had demonstrated hypercalciuria in hypophysectomized patients receiving a preparation of sheep prolactin. BECK et al. (1964 a, b) found that in pituitary dwarfs, receiving a highly purified preparation of ovine prolactin (PLR) prepared by a modification of the method of REISFELD, TONG, RICKES, BRINK and STEELMAN (1961), this preparation induced many of the metabolic changes that occur when these same individuals receive human growth hormone (MCGARRY and BECK, 1961). BECK et al. also found that their purified ovine prolactin induced a decrease in the blood urea nitrogen levels, potassium and phosphorus retention, and hypercalciuria.

MCGARRY, RUBINSTEIN and BECK (1968) extended these investigations. Their approach was based on comparison of metabolic effects, since, although immunologically STH and LTH might or might not differ, this has no bearing on their physiological task. Even the fact that, in a certain test, purified LTH fails to give a STH effect, or *vice versa*, does not attest to its task in the complicated mechanism of lactation. They feel that, chemically, STH and LTH are different materials, but their effects may overlap. A comparison of all the known biochemical and physiological effects of STH and LTH preparations shed light on the question in how far they are physiologically identical and can replace each other. They summarize the biochemical, immunological and physiological actions of STH and LTH preparations from different species on humans. The main biochemical effects of human STH were: decrease in blood sugar, decrease in free fatty acids (FFA) which later increased above the initial level, decrease in urinary and blood urea nitrogen. These effects of human STH were closely reproduced by ovine LTH, but not by other non-primate LTH or STH. In spite of the similarity of action of human STH and ovine LTH, no immunological cross-reaction was discovered either for the intact molecules or for their peptic, chymotryptic or papaic digests.

Apostolakis (1964) found that highly purified human LTH gave a very limited STH effect. The question may be raised whether, during the purification, certain physiologically reactive sites are not destroyed or polymerized. If we imagine STH and LTH as proteins with many active sites, rather than one active molecule with many possible actions, most of the contradictory findings in literature can be easily explained. If we accept this view, then LTH would appear to be a more or less anomalous STH, depending on the method of preparation and on the species. Such an anomaly—natural or artificial—might, of course, cause new effects and we should not expect all the physiological and biochemical properties of LTH to be present in STH. There is no doubt that all human STH preparations contain LTH activity, as exemplified by biological assay (Kovacic, 1962 a) and immunoassay (Laron and Assa, 1964; Laron, 1965). Its activity in promoting lactation in nursing women has been demonstrated by Lyons, Li, Ahmad and Rice-Wray (1968). Its failure to promote crop sac development has been shown by Forsyth (1967) and by Cowie, Hartmann and Turvey (1969).

Peake, McKeel, Jarett and Daughaday (1969) described a human pituitary tumor which contained LTH only, but no STH. This observations was extended by RIMOIN et al. (1968) who described an isolated deficiency of human growth hormone in 7 female ateliotic dwarfs. They had normal pregnancies and lactated normally in the post-partum period. One such female was studied during the last trimester and the early post-partum period of a normal pregnancy. Plasma STH could not be detected during the period of post-partum lactation when measured following insulin-induced hypoglycemia, arginine infusion and suckling. Since STH was not detectable during active lactation, neither in the basal state nor following provocative stimuli, the human growth hormone molecule, or at least those parts of it which are immunologically active and involved in growth promotion, do not seem to be necessary for post-partum lactation (PASTEELS, 1967).

Although the biochemical separation of the growth-promoting and prolactin activities of human and monkey growth hormone has not been accomplished, several lines of evidence suggest that these two activities reside in different molecules in primates, as they do in other species. CHEN and WILHELMI (1964) found that the ratio of the two activities in a variety of human STH preparations differed more than was to be expected if they were associated with the same molecule. ROTH, GLICK, CUATRECASAS and HOLLANDER (1967) found that serum LTH activity does not parallel the marked rise in immunoreactive serum STH observed following insulin-induced hypoglycemia, and that it is normal in acromegalics. Moreover, prolactin activity is elevated in the plasma of normal lactating women and in women with galactorrhea not associated with acromegaly—two states in which STH levels are not elevated (CANFIELD and BATES, 1965). Evidence derived from the above study of women with an isolated deficiency in human growth hormone supports and extends these observations.

One of the most interesting findings in regard to the biologic properties of STH in experimental animals is that human STH (HGH) possesses the various characteristics previously demonstrated by ovine lactogenic hormone: 1) it promotes pigeon-crop sac growth in both local and systemic tests; 2) it stimulates functional hypertrophy of corpora lutea cells, as shown by histologic study of the ovary and by the induction of deciduoma reactions; 3) it synergizes with the ovarian hormones to induce ductal and alveolar mammary growth; and 4) it induces localized milk secretion. However, these activities of HGH reach only 10 to 20 per cent of the extent of those of the animal lactogenic hormone. Nevertheless, no matter how we modify HGH, it still exercises these lactogenic activities side by side with the growth-promoting activity as long as the latter persists. Moreover, if the growth-promoting activity is destroyed, the lactogenic activities are abolished as well. There can, therefore, be no doubt that HGH has intrinsic lactogenic properties. Still, no similarity in chemical structure is evident when gross structures of HGH and ovine lactogenic hormone are compared. It seems that both hormones have an active core in common.

In summary, the similarity between human STH and LTH on the one hand, and the LTH activity of the human STH preparations on the other, raises serious doubts as to their purity and method of preparation. The first step in resolving this problem must be the preparation of crystallized protein hormones.

Chapter 5

Gonadotropins and Sex Steroids

Estrogen and progesterone are important for mammogenesis and have been shown to play a role in colostrogenesis as well (De Jongh, 1932). On the other hand, they are of secondary importance for the maintenance of lactation, i. e. lactogenesis and galactopoiesis. The gonadotropins are important in so far as they regulate the ovarian function. In some animals, such as rats (Evans, Simpson, Lyons and Turpeinen, 1941; Everett, 1944), mice (Dresel, 1935), and probably the ferret (Donovan, 1963 a, b), prolactin must be included among the gonadotropins, along with FSH and LH, on account of its luteotropic action in these species.

It is now well established that the stimulus for lactogenesis is supplied by the anterior pituitary (Folley, 1956; Lyons, 1958). The ovarian hormones are believed to be involved in the control of the anterior pituitary factors, or lactogenic complex, of which LTH and STH seem to be the most important components (Folley, 1956). Milk is present in the mammary gland during the second half of pregnancy of animals and man, but copious milk secretion does not set in until parturition. NELSON (1936) hypothesized that the high levels of estrogen during the second half of pregnancy inhibited both anterior pituitary release of prolactin and mammary tissue response to this hormone. This theory could not be proved for all species. MEITES and TURNER (1942 a, b; 1948), whose theory is based on determinations of the prolactin content of the pituitary, found that estrogen induces the secretion of prolactin from the anterior pituitary, thereby bringing about lactogenesis, whereas progesterone inhibits the action of estrogen, so that lactation is induced by the fall in the body level of progesterone occurring at parturition and the estrogen surge at term.

As to the question whether the ratio of progesterone to estrogen, or the absolute amounts of these hormones, are the critical factors in mammogenesis, it seems clear now—on the basis of thorough experiments carried out in the guinea-pig (BENSON, COWIE, COX and GOLDZWEIG, 1957)—that the absolute amounts of estrogen and progesterone determine the growth response of the mammary gland. Recent experiments performed in the goat give further proof of the secondary importance of the ovary in mammogenesis, since full mammary development could be obtained by regular milking of ovariectomized animals (COWIE, KNAGGS, TINDAL and TURVEY, 1968). It may be argued that in these experiments adrenal steroidogenesis could have replaced the missing ovarian hormones. This would also be consistent with the fact that male goats have sometimes been found to produce milk. Then, too, experiments performed with adrenalectomized-ovariectomized-hypophysectomized rats show that the ovaries and adrenals are of secondary importance for maintenance of mammary structure in this species, as measured by DNA (HAHN and TURNER, 1967).

The integrity of the ovary is not essential for the maintenance of lactation (galactopoiesis), since it depends on STH and TSH, and ovariectomy has no noxious effect on lactation (KURAMITSU and LOEB, 1921; DE JONGH, 1932; FOLLEY and KON, 1938; FLUX, 1955). In the intact rat very low doses of estrogen inhibit milk secretion, while in the ovariectomized rat much larger doses are required (COWIE, 1966). It is believed that ovarian progesterone synergizes with exogenous estrogen in the intact animal, the combination blocking the action of the lactogenic complex on mammary tissue (COWIE, 1966). Progesterone alone has no effect on milk secretion (FOLLEY, 1952 a, b), save in the adrenalectomized animal (FLUX, 1955), in which it may compensate for lack of adrenal steroids. The inhibitory action of progesterone would, therefore, become apparent only in combination with estrogen (FAURET, 1941; WALKER and MATTHEWS, 1949; COWIE, FOLLEY, MALPRESS and RICHARDSON, 1952; MEITES and SGOURIS, 1954).

The mechanism of the inhibitory action of estrogen + progesterone must, at least partially, be attributed to a direct effect on the mammary parenchyma (DESCLIN, 1950; MEITES and SGOURIS, 1953). *In vitro* experiments with dog mammary tissue, on the other hand, showed a dual effect of the estrogen-progesterone combination: very low levels proved to be stimulatory for lactogenesis, while higher levels exerted an inhibitory influence (BARNAWELL, 1967).

To sum up, it would seem that the gonadotropins and their secondary steroids—estrogen and progesterone—are important for the development of the mammary glands in the normal female. They are, however, not required for active lactation.

Chapter 6

Adrenocorticotropin and Corticosteroids

It has long been established that adrenal cortical hormones, as well as prolactin, are essential for maintenance of lactation. In hypophysectomized rats, guinea pigs or goats, these two hormones were shown to constitute the minimal requirements for milk secretion. For many years, the adrenal cortical hormones were considered by some workers—but not by FOLLEY (1956) and COWIE and FOLLEY (1961)—to have a permissive, and hence a relatively passive, role, whereas prolactin was believed to be the primary stimulator of lactation (MEITES, 1959 a).

Early evidence on the role of the adrenal in mammogenesis is conflicting (FOLLEY, 1952 a), and the whole question has been experimentally re-examined during the last fifteen years. The 11-oxygenated corticoids exert no mammogenic effect on the ovariectomized mouse, and they block the duct-developing effects of estrogen. On the other hand, 11-desoxycorticosterone synergizes the duct-developing action of estrogen in the spayed mouse (FLUX, 1954). ACTH has no mammogenic effect in intact mice (FLUX and MUNFORD, 1957), while hydrocortisone acetate is mammogenic at low doses and without effect at higher doses (MUNFORD, 1957). The rat responds differently to the corticoids, since glucocorticoids have been reported to be both mammo- and lactogenic in virgin animals (SELYE, 1954; JOHNSON and MEITES, 1955).

As pointed out by AHRÉN and JACOBSOHN (1957), the effect of corticoids on the mammary gland is conditioned by the endocrine state of the animal. They found that, in the hypophysectomized-ovariectomized rat, cortisone was unable to induce normal growth, even after addition of ovarian hormones. In the non-hypophysectomized spayed rat cortisone was lactogenic, its mammogenic effect becoming apparent only after addition of ovarian hormones. These authors consider that corticoids are mammo- and lactogenic only in conditions of homeostatic balance, while they may inhibit or alter mammogenesis in higher doses where they induce metabolic imbalance. In hypophysectomized-ovariectomized-adrenalectomized ("triply operated") rats, full duct development can be obtained when a corticoid is added to a combination of STH + estrogen (LYONS, LI and JOHNSON, 1958). In these rats, lobulo-alveolar development can be achieved by administering estrogen + progesterone + LTH + STH, provided that saline is supplied for drinking (LYONS, LI, COLE and JOHNSON, 1953).

Research during the past few years seems to indicate that ACTH and the adrenal cortex play as active a role in initiating and maintaining lactation as does prolactin itself. Thus, in intact pregnant rats (TALWALKER, NICOLL and MEITES, 1961) and in mice (NANDI and BERN, 1961), adrenal cortical hormones elicited lactation, whereas prolactin alone was ineffective. This suggests that the availability of adrenal cortical hormones rather than prolactin is the limiting factor in inducing lactation during pregnancy in these species; presumably, sufficient prolactin is already present during gestation. BALDWIN and MARTIN (1968 a, b) showed that both prolactin and cortisol are required for casein synthesis during lactation. An adrenal glucocorticoid hormone (9-fluoro-prednisolone acetate) was also demonstrated to induce lactation in pregnant cows (TUCKER and MEITES, 1965); prolactin alone was not tested. In pregnant rabbits, either cortisol acetate or prolactin could bring about lactation, and a combination of the two hormones was more effective than either was alone

(MEITES, HOPKINS and TALWALKER, 1963). This suggests that, in the pregnant rabbit, both types of hormones are limiting factors in initiation of lactation. It remains to be determined whether the rat and mouse are more representative of the mammalian species than is the rabbit in regard to the hormonal requirements for production of milk secretion during gestation.

BEN-DAVID and SULMAN (1969) have shown that acute adrenalectomy affects only slightly the full lobulo-alveolar development induced by high doses of perphenazine. WELLER, DANON and SULMAN (unpublished data) confirmed this finding. It was found that chronic adrenalectomy (for two weeks or longer) is clearly detrimental to the mammotropic effect of perphenazine. Whether chronic adrenalectomy impairs perphenazine-induced mammogenesis only on account of chronic deficiency of corticosteroids at body and mammary tissue level, or whether tropin disbalance at pituitary level also plays a role, can only be determined by future research. Preliminary experiments point to the validity of the second hypothesis. Many drugs, non-specific agents, stress interventions which have been shown to initiate mammary secretion in rats, rabbits, or other species, apparently do so because they elicit release of both prolactin and ACTH from the adenohypophysis (NICOLL, TALWALKER and MEITES, 1960; MEITES, 1961 a, b; MEITES, NICOLL and TALWALKER, 1963). In our rat strain, however, we have been unable to confirm the mammotropic and lactogenic effects of such agents or stress (WELLER and SULMAN, unpublished data).

The primary importance of pituitary integrity for the action of adrenal steroids on the mammary gland is emphasized by the findings of HAHN and TURNER (1967), according to which adrenalectomy has negligible effect on the DNA content of the mammary glands of hypophysectomized-ovariectomized rats. In these experiments, striking differences were detected only between the hypophysectomized and non-hypophysectomized groups, irrespective of the additional gland excisions. In the intact pregnant rat an optimal level of corticosterone was found which is able to induce maximum increases in the DNA and RNA titers of the mammary tissue; lower and higher levels show a reduced effect on DNA and RNA (KUMARESAN, ANDERSON and TURNER, 1967).

The role of the adrenal in galactopoiesis is of admitted importance, but opinions are at variance as to whether it is the gluco- or the mineralcorticoids which are primarily involved. Glucocorticoids are considered critical by GAUNT, EVERSOLE and KENDALL (1942), while FOLLEY and COWIE (1944) and COWIE and FOLLEY (1947) argue that mineralocorticoids are crucial. Later experiments (COWIE, 1952; COWIE and TINDALL, 1955) suggest that both types of corticoids are necessary for optimal milk secretion. In general, the same conclusions were drawn from experiments with adrenalectomized goats, though in these studies mineralocorticoids (DOC) were found to play a more critical role than glucocorticoids, e. g. cortisone (COWIE and TINDAL, 1958).

Since the ovary may compensate for adrenal steroids by way of progesterone production (FLUX, 1955), it must be realized that differences in the degree of ovarian contribution may be due to strain differences in rats. This holds specially with regard to the relative importance of gluco- or mineralocorticoids for galactopoiesis in adrenalectomized animals (COWIE and FOLLEY, 1961). In intact cows, ACTH and cortisone depress lactation (COTES, CRICHTON, FOLLEY and YOUNG, 1949; FLUX, FOLLEY and ROWLAND, 1954; SHAW, CHUNG and BUNDING, 1955). In the rat, ACTH and cortisone have galactopoietic effects (JOHNSON and MEITES, 1958); large doses of cortisone, however, inhibit milk secretion in this animal (MERCIER-PAROT, 1955).

There is growing evidence that increased amounts of prolactin, corticotropin and adrenal cortical hormones are released at about the time of parturition (COWIE and FOLLEY, 1961; MEITES, 1959 a), and they all serve to initiate lactation. After parturition, the milking or suckling stimulus results in a realease of LTH, ACTH, and corticosteroids (COWIE and FOLLEY, 1961; MEITES, 1959 b), and these act to maintain milk secretion. Other hormones (STH, thyroid hormones, insulin, and parathormone) are necessary to elicit maximum production of milk, but are not essential for milk secretion. However, the importance of adrenal steroids in galactopoiesis may lie, at least in part, in their permissive action as concerns prolactin activity (BARNAWELL, 1967), as well as in their more general metabolic effects (AHRÉN and JACOBSOHN, 1957). Thus, for instance, they increase the formation of mammary enzymes (HEITZMAN, 1968).

In conclusion it should be said that the participation of all corticoid hormones is required for lactation. The presence of still unknown corticosteroids, e. g. Hartman's "Corticolactin" in the amorphic fraction of cortical extracts, cannot yet be excluded.

Chapter 7

Thyrotropin and Thyroid Hormones

The thyroid would seem to influence mammogenesis by its general metabolic effect (JACOBSOHN, 1958, 1959). A certain degree of hypothyroidism enhances alveolar development in the rat, but it has an inhibitory effect on mammogenesis in the mouse (FOLLEY, 1956). Treating rats with propylthiouracil (an antithyroid drug) stimulates mammary growth (BEN-DAVID, DIKSTEIN and SULMAN, 1966). Mammogenesis can be elicited in the hypophysectomized-adrenalectomized-thyroidectomized rat by giving ovarian hormones + prolactin, STH and hydrocortisone, without the addition of thyroid hormones (CHEN, JOHNSON, LYONS, LI and COLE, 1955).

The thyroid does not seem to be essential for lactogenesis and galactopoiesis either, though its absence may deleteriously affect the intensity and duration of milk secretion (FOLLEY, 1952 b). Treating lactating rats with propylthiouracil depresses milk yield (BEN-DAVID, DIKSTEIN and SULMAN, 1966). That thyroid activity is enhanced during lactation is supported by histologic (RACADOT, 1957) and biochemical (GROSVENOR and TURNER, 1958 b) studies. GRAHAM (1934) showed that, by feeding dried thyroid gland or by injecting thyroxine, the milk yield of cows could be increased. Some years later it was found that certain iodinated proteins possessed thyroxine-like activity when given in the feed (COWIE, 1956). It was shown that thyroxine is galactopoietic when fed to lactating cows in daily doses of about 100 mg (BAILEY, BARTLETT and FOLLEY, 1949). As regards 3, 5, 3'-triiodothyronine, it was ineffective when orally administered to cows, probably because it is inactivated in the rumen. On the other hand, it is more galactopoietic than thyroxine when administered subcutaneously (BARTLETT, BURT, FOLLEY and ROWLAND, 1954). There is a wide variation in the degree of stimulation of milk production resulting from the administration of thyroxine or of thyroid-active substances; on the whole, a better response is obtained during the decline of lactation than at the peak or end of lactation (HERMAN, GRAHAM and TURNER, 1938; RALSTON, COWSERT, RAGSDALE, HERMAN and TURNER, 1940; BLAXTER, 1945), but it is not possible to make a good producer out of a poor producer by feeding thyroprotein (THOMAS, 1953). Moreover, it has been found that the use of thyroxine or thyroid-active substances during the first to third lactations is inadvisable, for the boost in yield obtained is largely cancelled out by a shortening of the lactation period (THOMAS and MOORE, 1953; COWIE, 1956).

The best results are obtained when the thyroid-active substances are administered for relatively short periods, i. e. not exceeding two months. Short-term administration at favorable periods can result in dramatic increases (REINEKE, 1942). Unfortunately, because it takes some time for the animal's own thyroid to recover from its quiescent state, when administration ends there is often an equally spectacular decrease in yield, which is due to the sudden fall in the thyroxine level in the blood (SWANSON, 1954; COWIE, 1956).

The careful administration of thyroid-active substances to dairy cows has no deleterious effect on the health and reproductive ability of the cows (LEECH and BAILEY, 1953). Unfortunately, there is only a net gain in yield of about 3 per cent, so that the practical application of the procedure seems to be limited (COWIE and FOLLEY, 1961).

Thyrotropin, as was to be expected, increases milk yield by increasing the output of thyroxine from the thyroid (COWIE, 1956). It is unlikely that thyroxine has a specific effect on the alveolar cells, so that its effect on galactopoiesis should probably be ascribed to its general metabolic action (MEITES, 1961 a, b).

Summarizing, it seems that TSH and its secondary hormones are not required for the first stages of lactation. Similarly, thyroxine and thiouracil do not affect hypothalamic PIF (CHEN and MEITES, 1969).

Chapter 8

Oxytocin and Vasopressin

There can no longer be any doubt that oxytocin controls milk ejection, and it is generally accepted that milk ejection is mediated mainly by a humoral mechanism—the release of oxytocin in response to suckling (BISSET, 1968). It is also clear that suckling not only initiates the milk ejection reflex, but it is also necessary for maintenance of milk secretion in the lactating rat. The salient point would then be the problem whether synthesis of milk by the cells of the alveolar epithelium and the passage of milk from the cytoplasm of the alveolar epithelial cells into the alveolar lumen are dependent on oxytocin as assumed by SELYE (1934 a, b) and SELYE, COLLIP and THOMPSON (1934). It was concluded years ago that involution of the mammary glands in the rat after weaning is due not to accumulation of secreted milk, but to withdrawal of the suckling stimulus, and it was suggested that suckling stimulated also the secretion of prolactin from the anterior pituitary. Thus, PETERSEN (1944) as well as BENSON and FOLLEY (1956), BENSON (1960) and BENSON, COWIE and TINDAL (1958) advanced the idea that the oxytocin released by suckling constitutes a stimulus for prolactin release. This hypothesis was based on their success in retarding the involution of the lactating mammary gland by injections of oxytocin. Their theory was given weighty support by their finding that oxytocin stimulated mammary development only in intact—but not in hypophysectomized—mothers. Thus, the theory was developed that oxytocin, released from the nerve endings in the neurohypophysis by the milk-ejection reflex, reaches the adenohypophysis via the hypophysial portal vessels and stimulates it to release prolactin. This hypothesis is supported by the studies of MCCANN, MACK and GALE (1959) who found that rats, bearing hypothalamic lesions in the supraoptic tract, suffered from a block of milk secretion and ejection which could be overcome by oxytocin.

MEITES and his group (MEITES, 1959 a; MEITES and NICOLL, 1959; MEITES and HOPKINS, 1961) demonstrated that a combination of prolactin with oxytocin is more effective than prolactin alone in retarding mammary involution, but they felt that two independent mechanisms controlled the release of oxytocin and prolactin by suckling, and that oxytocin retarded mammary involution not by releasing prolactin, but by a direct local action on the epithel of the mammary gland. The emptying of the alveoli by the action of oxytocin on the myoepithelium reduces the pressure of accumulated secretion against the epithelial cells and the blood capillaries, thus permitting synthesis of milk to proceed under the influence of prolactin (MEITES, NICOLL and TALWALKER, 1960; MEITES, NICOLL and TALWALKER, 1960).

Experiments by GALA and REECE (1965 a) and by our group (MISHKINSKY, DANON and SULMAN, unpublished results) both *in vivo* and *in vitro,* favor the concept that oxytocin can release prolactin from the rat anterior pituitary, and that any discrepancies are due to the fact that the effect is dose-dependent: Small doses of oxytocin stimulate prolactin release, but larger doses have in inhibiting effect. This aspect is still under intensive research, requiring the use of long-acting oxytocin preparations or repeated daily injections. Nevertheless, it is clear that whether or not oxytocin stimulates indirectly milk secretion by releasing prolactin, its direct action on the mammary gland is the cause of ejection of pre-formed milk from the alveoli.

For many years, the pharmacological activities of oxytocin and vasopressin were found to be overlapping. This was due to incomplete separation of the two hormones, and it was not until 1941 that ELY and PETERSON concluded that oxytocin was the natural hormone responsible for milk ejection, while the effect of vasopressin on the mammary gland was due to contamination by oxytocin. Moreover, a third factor was thought to be involved in the posterior pituitary; it was defined by TURNER, PIERCE and DU VIGNEAUD (1951) as arginine-vasopressin and found to have about one-fifth the potency of oxytocin when tested for milk-ejecting activity. Thus, it was concluded that milk-ejecting activity, like oxytocin activity, is an intrinsic property of vasopressin. It seems that although vasopressin may be released during suckling, it is the release of oxytocin which brings about milk ejection. The intravenous injection of this hormone will produce milk-ejection ("let-down") within a minute. This has as yet no practical application in dairy husbandry. Therapeutically it may be useful in sows which fail to eject milk after farrowing.

Some doubt has been expressed on whether oxytocin is the milk ejection factor in primiparous women. It is certainly active in pharmacological doses (2—4 IU intranasally) and increases milk yield. The milk-ejection reflex may be excited by various kinds of conditioned stimuli, but it is also subject to inhibition by emotionally stressful situations, in women as well as in cows (Fig. 13).

The reaction of oxytocin release in response to conditioned stimuli associated with machine milking in the cow was recently studied by CLEVERLEY (1967). Each cow was accustomed to a strict milking routine, which included the presence of the experimenter, for at least 10 days before blood sampling. Approach and entry of the milker stimulated the release of oxytocin. These results provide, for the first time, direct experimental evidence that the release of oxytocin can be conditioned to auditory and visual stimuli associated with the milking process. Similarly, washing the udder and foremilking also stimulated the release of oxytocin. Injections of 100 MU synthetic oxytocin into the external jugular vein near milking time elicited a rise in intra-mammary pressure equivalent to the natural ejection recorded during milking, with a latency of 40—50 sec. Recording was continued for 1 hr. after milking or oxytocin injection, and during this time the elevated intramammary pressure was maintained. These experiments show that oxytocin is released transiently and at low levels during milking in the cow. It can be released in response to conditioned stimuli, and/or direct physical stimuli applied to the udder.

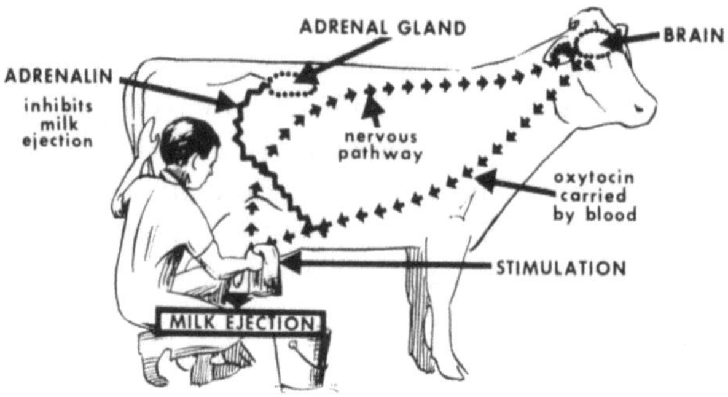

Fig. 13. Effect of oxytocin on milk yield: Milk ejection ("let-down") by oxytocin and milk retention by adrenaline following stress, pain or fright (by courtesy of F. P. Natzke, A. M. Meek and D. K. Bandler, 1967, N. Y. State College of Agriculture, Cornell University, Ithaca, N. Y.). — 1. Udder washing stimulates nerves which carry the message to the brain to release oxytocin to the blood. Oxytocin reaching the udder causes milk ejection. 2. Pain or fright releases adrenalin which blocks the action of oxytocin (milk ejection)

In conclusion, it emerges that oxytocin (and vasopressin) can increase milk output, the main mechanism being the emptying of the myoepithelial elements of the mammary gland. Its central effect on the PIF or PRF is doubtful because it seems to be dose-dependent. This requires further study, in view of the fact that the posterior pituitary releases oxytocin during nursing, suckling and milking by direct stimuli as well as by conditioned stimuli.

Chapter 9

Parathormone and Calcitonin

Not many data exist on the effect of the parathyroid on lactation. Experiments by Smith, Stott and Walker (1957) were carried out on parathyroid and thyroidectomized goats. They found an inhibition of lactation.

Parathormone appears to be necessary for full lactation, and its absence can result in interference with the calcium-concentrating mechanism of the mammary gland (COWIE and FOLLEY, 1961). No systematic study of the parathyroid hormone and/or calcitonin and their influence on mammogenesis or lactation has been carried out. There exists, however, one important clinical observation during lactation: Many species, especially cows, dogs and pigs, can develop lactation tetany, as their demand of calcium for milk production is excessively high. The condition is rare in women but, if observed, it points to a parathyroid deficiency.

The effect of calcitonin on the calcium and phosphorus levels of lactating cows has been studied by BARLET (1967). Injecting 20 RMC Units, he obtained a 20% reduction in blood calcium level and a 35% decrease in blood phosphates. These results indicate the possible importance of thyroid calcitonin in the development of lactation tetany.

Chapter 10

Insulin and Glucagon

Insulin does not seem to affect lactation *in vivo* unless given in relatively large doses. In this case the milk yield is reduced, apparently due to the severe hypoglycemia which may affect the supply of galactose. This detrimental effect was abolished by simultaneous administration of insulin and glucose (KRONFELD, MAYER, ROBERTSON and RAGGI, 1963). Smaller doses, which do not lower blood sugar levels substantially, were tried in guinea pigs (0.2 IU insulin/day) for 14 days after parturition. This treatment improved the milk yield and induced increased growth of the offspring (CORDELLI, CASTELLI and GORELLI, 1960).

Small doses of insulin, if given together with ovarian hormones, can apparently induce some development of the mammary ducts in hypophysectomized rats; presumably insulin corrects some of the metabolic deficiencies which result from pituitary ablation (AHRÉN and JACOBSOHN, 1956; JACOBSOHN, 1961). However, it has not been demonstrated that injections of insulin alone can elicit mammary growth in either intact or hypophysectomized animals. The stimulating effect of insulin on mammary tissue *in vitro* has recently been shown by TOPPER (unpublished results of Laurention Hormone Conference 1969).

Glucagon has not been studied with regard to its effect on lactation. It may, however, be safely assumed that small quantities stimulate lactation due to glucose mobilization. On the other hand, larger doses have been shown by PALOYAN et al. (1965, 1967) to induce hypocalcemia, which would be detrimental to milk yield.

Chapter 11

Pineal Gland

Until recently very little was known about the effect of the pineal gland on mammary development and activity. KITAY and ALTSCHULE (1954), in their review of the literature on the physiology of the pineal gland, quote three references to the "galactogenic" action of pineal extracts, which in fact are reports of milk ejection occurring after the administration of these extracts. It is now known that they may contain arginine-vasotocin (PAVEL and PETRESCU, 1966). The question of a possible pineal endocrine effect on milk secretion was considered by THIEBLOT (1954), who concluded that, while no direct evidence of any effect on milk secretion was available, there was, however, proof that the pineal can inhibit the pituitary and its luteotropic activity, and that further studies on the pineal in relation to milk secretion were called for.

In a study carried out by MISHKINSKY, NIR, LAJTOS and SULMAN (1966), female rats were pinealectomized 21 days after birth. When 26, 35, 50 and 100 days old, their inguinal mammary glands were excised and the degree of mammary development evaluated (MTI). Inguinal mammary glands of normal and sham-operated female rats were also investigated for comparison. As in the controls, the maximal development in the pinealectomized animals reached MTI stage 2, indicating that no change had occurred in their mammary development. In addition, in order to promote development of the mammary gland, 21-day-old rats weighing 40 ± 4 g, and adult female rats weighing 200 g, were primed for ten days with 10 µg estradiol/kg/day and

8 μg estradiol/rat/day respectively and then pinealectomized. Their mammary glands were examined six days later and again the MTI (stages 1—2) did not differ from that of the normal and sham-operated controls subjected to the same treatment.

As pinealectomy did not enhance mammary development nor the release of mammogenic or lactogenic hormone in female rats between the ages of 21 and 100 days, it became obvious that the pineal has no inhibitory effect on the development of the mammary gland.

In a further study attention was concentrated on pineal control of lactogenesis and maintenance of milk secretion in mother rats after parturition. This was done by comparing the milk yields of pinealectomized rats with those of untreated and sham-operate controls (NIR, MISHKINSKY, ESHCHAR and SULMAN, 1968). Female rats were pinealectomized when 21 days old and mated when four to six months old. Fourteen, 16, 18 and 20 days post-partum their milk yields were determined by the increase in the litter weights during suckling (after the litters had been adjusted to six pups each). The milk yields of the pinealectomized mothers were about 13 per cent lower (except on the 20th day) than those of the sham-operated controls, but not lower than those of the intact controls.

Our results suggest that removal of the pineal causes some depression of lactation in the rat, yet the differences in yields between the groups did not reach the accepted level of statistical significance ($P > 0.05$). Moreover, interpretation of the effects of pinealectomy was complicated by the unexpected tendency of the sham-operated controls to have higher milk yields than the intact controls. Further research is clearly necessary in order to determine the true significance of these effects before the role of the pineal in lactation can be established.

Chapter 12

Placental and Amniotic Fluid

In recent years it has been claimed that the human placenta, besides its gonadotropins, secretes also a mammotropic hormone, which has been called "Purified Placental Protein" (PPP) by FLORINI, TONELLI, BREVER, COPPOLA, RINGLER and BELL (1966), "Human Placental Lactogen" (HPL) by JOSIMOVICH and MACLAREN (1962),

"Chorionic Growth Hormone Prolactin" (CGP) by KAPLAN and GRUMBACH (1965 a, b), and "Placental Protein" (PP) by FRIESEN (1965). FLORINI's 70% pure PPP was reported (1966) to contain 2—4 IU/mg of prolactin, as assayed by the pigeon crop sac method.

Although EHRHARDT demonstrated prolactin activity in extracts of human placenta as early as 1936, considerable doubt was raised whether there exists a human placental lactogenic hormone (JOSIMOVICH and MACLAREN, 1962). The existence of a prolactin-like substance (HPL) with immunological and lactogenic properties *in vivo* is now well established (FLORINI et al., 1966; FORSYTH, 1967), and its effect *in vitro* was shown by TURKINGTON and TOPPER (1966). It has neither hyperglycemic nor diabetogenic effect (KURCZ, 1967). HPL seems, therefore, to fulfill an essential function both in mammogenesis and colostrogenesis.

In a recent study by our group (MISHKINSKY, ZANBELMAN, KHAZEN and SULMAN, 1969) the mammotropic and prolactin-like effects of rat placenta homogenates, amniotic fluid and placental lactogen, as well as human placental lactogen and amniotic fluid, were measured by different assay-methods. Mammary development (MTI) was determined on days 2, 4, 7, 10, 14, 17 and 20 of pregnancy or pseudopregnancy (uterine-horn scratch).

Suspensions of rat placentas were prepared in normal saline by homogenizing placentas taken from pregnant rats at a later stage of pregnancy (16th day) with a cone-glass electric homogenizer soaked in ice. When removing the placenta, the amniotic fluid was collected and stored at —20° C. The amniotic fluid or placental suspensions were injected subcutaneously once daily, for five days, to groups of 20—24 virgin rats, weighing 200 ± 10 g, and primed with estradiol (8 μg/rat/day) for 10 days. The doses chosen were the equivalent of 1 or 4 placentas/rat/day, and 0.5 or 2.0 ml amniotic fluid/rat/day; 24 hours after the last injection the rats were decapitated and their MTI evaluated. Rat placental lactogen (RPL) was prepared according to the method of JOSIMOVICH and MACLAREN (1962) for preparation of HPL: 100 rat placentas were homogenized in 2 vol. of chilled tris-maleate buffer at pH 8.6 and brought gradually to pH 4.5, centrifuged, lyophilized and weighed (yield was 20—22 mg/placenta).

The crop sac test was performed according to NICOLL (1967), and statistical evaluations were made by the "rank sum" method for ordinal data described by WHITE (1952).

The MTI values of different stages of pregnancy and pseudopregnancy in the rat are presented in Fig. 14. The MTI of pregnant rats increases rapidly throughout pregnancy. However, no mammotropic development was noticed in the adult pseudopregnant rats, nor was any clear-cut correlation established between the MTI of virgin pseudopregnant rats and the weights of their uteri (Fig. 15), though both phenomena are prolactin-dependent, and the uterine weight curve partly resembles the ascending curve of prolactin-activity dur-

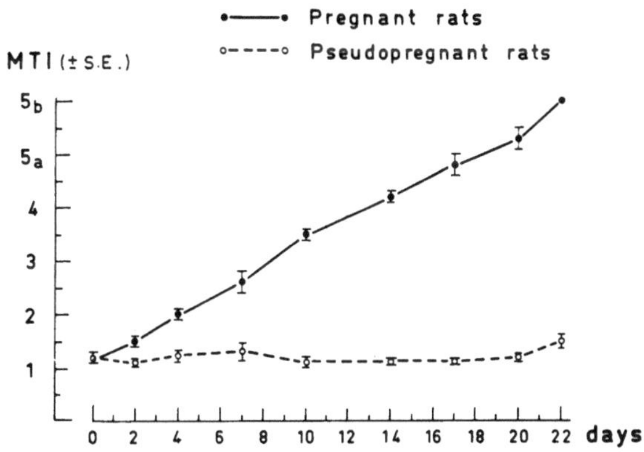

Fig. 14. Development of mammary gland in pregnant rats compared with lack of mammary differentiation in pseudopregnant rats

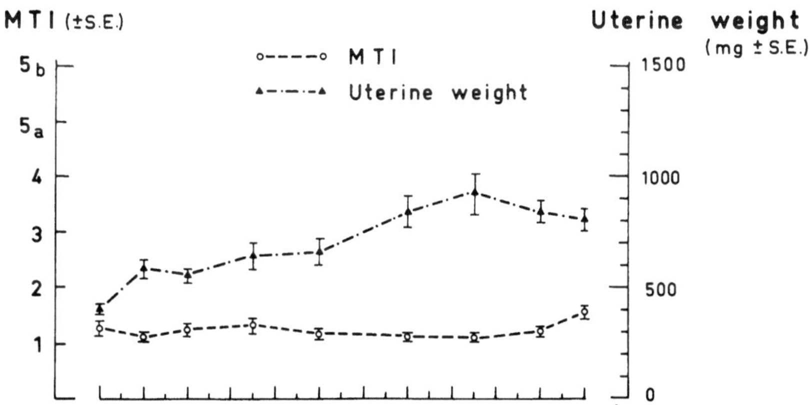

Fig. 15. Development of mammary gland in pseudopregnant virgin rats (lower curve) compared with decidual transformation and weight increase in uterus (upper curve)

ing pregnancy. Similar results were reported by DESJARDINS, PAAPE and TUCKER (1968).

The mammotropic activity of rat placental homogenates injected s.c. was evidenced only when the equivalent of 4 placentas was injected daily into each primed rat. It required this concentration for the difference between the MTI obtained after placenta homogenate injections and that of the controls to become significant ($P < 0.05$). Rat amniotic fluid had no evident mammotropic activity when injected s.c. into primed rats, nor when injected intradermally (0.6 ml native or 10 × concentrated) over the pigeon crop. Rat placental lactogen, however, was found to have a clear prolactin-like effect when measured by the pigeon crop sac proliferation test.

Human placental lactogen (HPL) was found to be mammotropic in all doses tested (1—100 mg/rat/day). In the crop gland test it yielded also a prolactin-like effect in doses of 0.015—1 mg. On the other hand, human amniotic fluid was found to have a moderate mammotropic effect when 0.5 or 2.0 ml were injected s.c. into primed rats, but was devoid of prolactin-like activity when injected intradermally (0.06 ml native or 10 × concentrated) over the pigeon crop. From these findings it emerged that rat and human amniotic fluids contain less than 0.01 IU/ml of prolactin or prolactin-like substance.

The study of HPL and RPL was undertaken by us because the experiments of WRENN, BITMAN, DE LAUDER and MENCH (1966) failed to prevent the decline in mammary development of pseudopregnant rats by various placental suspensions. Such experiments succeeded only when placental suspensions were injected into hypophysectomized-ovariectomized rats (RAY, AVERILL, LYONS and JOHNSON, 1955). Our findings agree with those of WRENN et al. (1966) and MATTHIES (1967), who found that the minimum effective dosage of rat fetus and placental extracts from the twelfth day of pregnancy, capable of stimulating the decidual response in hypophysectomized rats, were the equivalent of one placenta per day, injected s.c.

Summarizing, it is concluded that human and rat placenta contain a prolactin-like substance. The amniotic fluid of women contains only minute quantities of placental lactogen, and that of the rat does not contain any measurable amount at all (< 0.01 IU/ml).

Part II

Pharmacological Regulation of Lactation

Chapter 13

Hypothalamic Lactation Produced by Psychopharmaca

a) History

Tranquilizing drugs have been reported to interfere with a variety of endocrine functions, including lactation, in man and experimental animals. In 1954 WILKINS reported the development of gynecomastia following treatment with Rauwolfia alkaloids. WINNIK and TENNENBAUM (1955) showed that treatment with chlorpromazine (Largactil, Thorazine, Megaphen) might result in milk secretion in women. SULMAN and WINNIK (1956), and WINNIK and SULMAN (1956), studying this problem, described abnormal lactation in women and animals after treatment with chlorpromazine. Abnormal lactation was also reported to occur in women who received either reserpine or chlorpromazine (MARSHALL and LEIBERMAN, 1956; PLATT and SEARS, 1956). This abnormal lactation was confirmed in experimental animals as well. Treatment with reserpine resulted in milk secretion in rabbits (MEITES, 1957a, 1963; SAWYER, 1957; MEITES, NICOLL and TALWALKER, 1959, 1963) and in rats (DESCLIN, 1958; GRÖNROOS, KALLIOMÄKI, KEYRILÄINEN and MARJANEN, 1959). Administration of various phenothiazine derivatives also resulted in mammary differentiation and milk secretion in rats (KHAZAN, PRIMO, DANON, ASSAEL, SULMAN and WINNIK, 1962; TALWALKER, RATNER and MEITES, 1963; BEN-DAVID, DIKSTEIN and SULMAN, 1965). This endocrine disturbance is presumed to result from the action of tranquilizers on the reticular formation and hypothalamic regions which regulate the normal function of the pituitary (COURVOISIER, FOURNEL, DUCROT, KOLSKY and KOETSCHET, 1953; DELAY and DENIKER, 1953; PLUMER, EARL, SCHNEIDER, TRAPOLD and BARRETT, 1954). Thus the name "hypothalamic tranquilizers" has been given to these agents from which the conception of endocrine disturbances inducing lactation originated. They do not seriously affect "balanced tropins," i. e. tropins which are held in abeyance by the secondary hormone they produce, as shown in Fig. 16 (SULMAN, 1956 b). They do, however, seriously interfere with the function of the "unbalanced tropins" (LTH, MSH, ADH and oxytocin) as the latter do not produce secondary hormones which could counteract an overshoot produced by hypothalamic depression. Consequently, hypothalamic lactation results (Fig. 17) (SULMAN, 1959).

Many drugs and nonspecific agents have been shown to elicit mammary growth and secretion in estrogen-primed rats and rabbits if

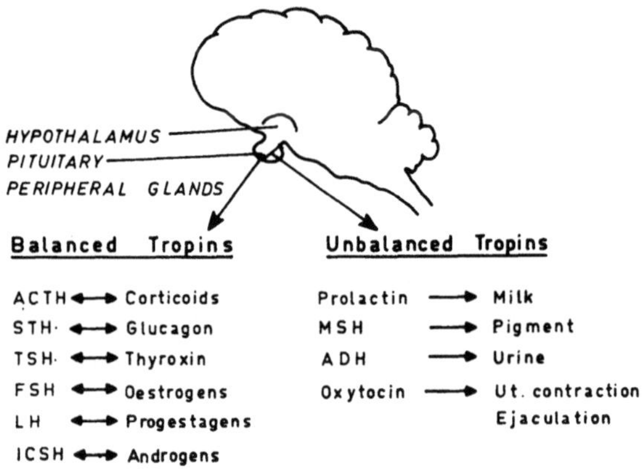

Fig. 16. Hypothalamic lactation: Balanced tropins have a positive and negative feedback mechanism. Unbalanced tropins react by overshoot when the hypothalamus is suppressed by phenothiazines

Hormonal reaction	Balanced tropins				Unbalanced hormones ↑
Week	1	2	3	4	Prolactin: Milk ↑
Hypothalamus	↓	↓	↓	↓	MSH: Pigment ↑
Pituitary	—	↓	↓	—	ADH: Urine ↓
Peripheral hormones	—	—	↓	—	Oxytocin: Ut. Contract. ↓ Ejaculation ↓

Fig. 17. Reaction to hormonal depression by phenothiazine derivatives: Balanced tropins overcome depression of hypothalamus by feedback. Unbalanced hormones have no peripheral feedback. This results in stimulation of prolactin and MSH and in suppression of ADH and oxytocin

the priming dose is set high, and to maintain secretory activity in post-partum rats after litter removal (MEITES, 1963; MEITES, NICOLL and TALWALKER, 1963). The effective drugs in rats include—in addition to chlorpromazine and reserpine—meprobamate, adrenaline, noradrenaline, serotonin, atropine, dibenamine, eserine, pilocarpine, amphetamine, morphine, 3-methylcholanthrene, 3,4-benzpyrene, 9,10-dimethyl-1,2-benzanthracene and 10% formaldehyde. Stress, such as cold, heat, restraint, and electrical stimulation also induce mammary growth. The above "extra-hypothalamic" psychopharmaca will only provoke lactation if estrogen priming has been too high, e. g. 10 µg/rat/day for 10 days. We have shown that meprobamate, hydroxyine, ethinamate, phenobarbital and benactyzine are devoid of any mammotropic effect in rats primed for 10 days with sub-threshold doses (8 µg/day) of estrogen (BEN-DAVID, DIKSTEIN and SULMAN, 1965). On the other hand, when high threshold doses of estrogen (10 µg/day) were used for priming, many non-specific stressful stimuli, as reported above, can initiate secretion by the mammary gland.

b) Dissociation of Mammotropic and Sedative Effects

In a trial to dissociate the sedative from the mammotropic property of hypothalamic tranquilizers (KHAZAN, BEN-DAVID, MISHKINSKY, KHAZEN and SULMAN, 1966), two ways of approach were chosen in assaying the mammotropic effect in rats: Biological dissociation by combined treatment with tranquilizers and energizers, controlled by behavioral and electroencephalographic studies; chemical dissociation by changes in the molecule of phenothiazine tranquilizers which would eliminate the sedative effect. This effect was gauged by prolongation of the barbital sleep induction time in mice, and lowering of the body temperature (BTD) in rats (GREIG and MAYBERRY, 1951; CHILD, SUTHERLAND and TOMICH, 1961, 1962). The BTD is a standard test for hypothalamic depression, since the thermoregulatory centers of the thalamus are connected with the emotional centers of the hypothalamus.

c) Biological Dissociation through Combined Treatment with Tranquilizers and Psychostimulants

KHAZAN, BEN-DAVID, MISHKINSKY, KHAZEN and SULMAN (1966) investigated whether the mammotropic property of a tranquilizer is independent of its sedative effect and resistant to changes of mood produced by psychostimulants which block the sedative-depressant effects.

Groups of 10 adult female rats, weighing 200 ($\pm$10) g, were used in each series of experiments. They were primed by daily s.c. injections of 8 µg estradioal in oil for a period of 10 days. From the 11th to the 15th day the following daily treatments were administered s.c.: (1) saline as control, (2) perphenazine HCl 2.5 mg/kg, (3) tetrabenazine methane sulphonate 5 mg/kg, (4) combined treatment of etryptamine 5 mg/kg and perphenazine·HCl 2.5 mg/kg, (5) combined treatment of imipramine·HCl 10 mg/kg with tetrabenazine methane sulphonate 5 mg/kg, (6) combined treatment of ephedrine·HCl 5 mg/kg with perphenazine 5 mg/kg, (7) etryptamine 5 mg/kg alone, (8) imipramine ·HCl 10 mg/kg alone, and (9) ephedrine 5 mg/kg alone. From the 11th to the 16th day, the EEG and overt behavior of the rats were evaluated three times daily (Fig. 18). On the 16th day the rats were sacrificed and their MTI determined.

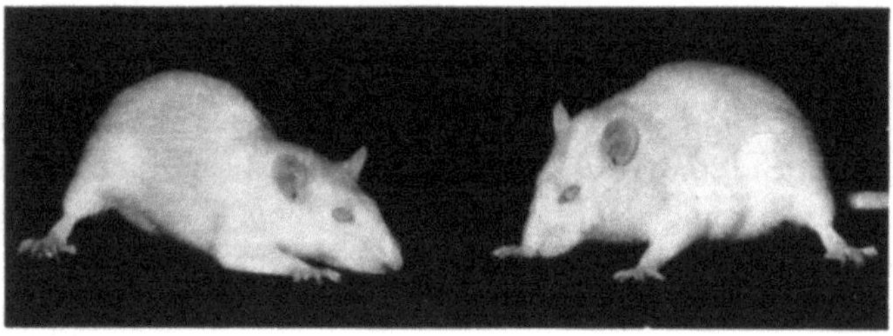

Fig. 18. Effect of perphenazine on overt behavior of rats after 5 days' treatment with: r: perphenazine 2.5 mg/kg s. c. alone—note tranquilizing effect; l: perphenazine 5 mg/kg s. c. combined with etryptamine 5 mg/kg s. c.—note alert behavior

The control rats showed normal proliferation of the mammary gland with an average MTI of 1—2. Perphenazine- and tetrabenazine-treated rats showed highly developed mammary glands (MTI 4—5). Combined treatment with etryptamine, imipramine or ephedrine, which blocked the sedative effects of perphenazine and tetrabenazine on the EEG and on overt behavior, nevertheless yielded a full mammotropic effect (MTI 4—5). The EEG and overt behavior were normal during the 5 days of observation (Fig. 19).

From the results obtained it can be concluded that the mammotropic and sedative effects of tranquilizers are not correlated, but are separate properties inherent in their molecules. The mammotropic properties are clearly independent and completely resistant to energizers which antagonize the tranquilizing effect.

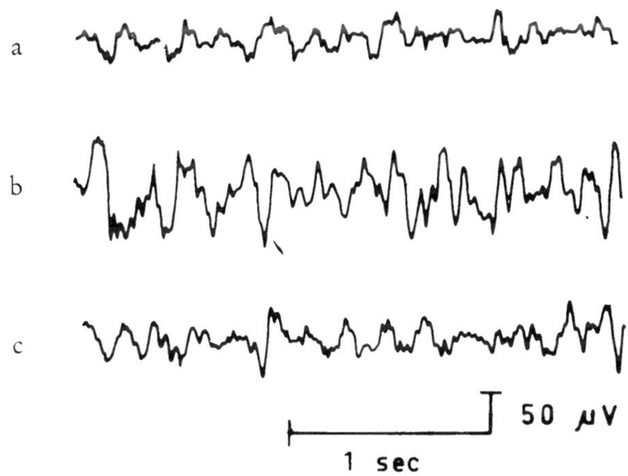

Fig. 19. Electrocortical tracings from unrestrained chronically implanted rats (a). Note the high voltage low frequency tracings induced by perphenazine (2.5 mg/kg) in (b), compared with the normal awake recording in (a). In (c) the desynchronizing action of etryptamine on the perphenazine effect is obvious

d) Chemical Dissociation by Molecular Changes of Tranquilizers

Khazan, Ben-David, Mishkinsky, Khazen and Sulman (1966) used chemical changes for dissociation to evaluate the tranquilizing and mammotropic effects of three typical substances: a phenothiazine sulfoxide (RP-8580), 2-chlorpromazine and 2-bromopromazine (Fig. 20). RP-8580 showed no tranquilizing, but strong mammotropic effects. Chlorpromazine showed similar tranquilizing effects as 2-bromopromazine, but was much less mammotropic. From these results it can be concluded that the mammotropic and sedative effects of tranquilizers are separate properties inherent in their molecules. They may have a partially overlapping affinity to hypothalamic receptors, but the tranquilizing and mammotropic effects are by no means coupled with each other. Lack of parallelism between tranquilizing and mammotropic effects of 15 reserpine derivatives was described by BEN-DAVID, KHAZEN, KHAZAN and SULMAN, 1968) (Chapter 16). BEN-DAVID (1968 a) recently found that persistent influence of a drug must be maintained in order to induce artificial hypothalamic lactation. This became evident when a single high dose

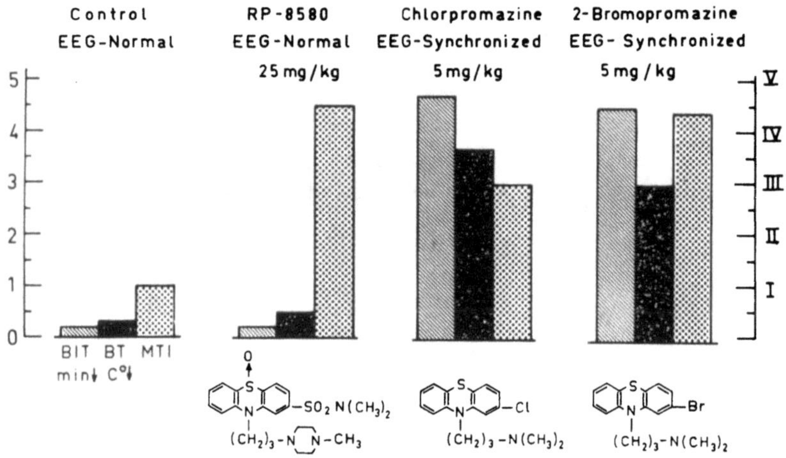

Fig. 20. Dissociation of barbiturate induction time (BIT), body temperature drop (BT) and mammotropic index (MTI), obtained with three different phenothiazine derivatives: RP-8580, yielding only high MTI; chlorpromazine, yielding high sedative reaction but low MTI; 2-bromopromazine, yielding high BIT and MTI but low BT

of fluphenazine-HCl (the short-acting form of the drug) failed to induce mammary differentiation comparable to that obtained by one injection of a high dose of fluphenazine enanthate (Chapter 14). The latter long-acting form of the drug converted the non-developed gland (control) into a well-developed gland with copious milk secretion. Thus it seems that the depressant activity of a phenothiazine drug on the hypothalamic prolactin-inhibiting factor (PIF)—which results in oversecretion of pituitary prolactin—can be accomplished only by continuous exposure to the drug for a minimum of 4 days.

e) Studies in Hypophysectomized Rats

Direct evidence that perphenazine stimulates pituitary prolactin secretion by suppressing the hypothalamic PIF is to be found in the *in vitro* work of DANON, DIKSTEIN and SULMAN (1963), who showed that perphenazine, added to pituitary-hypothalamus co-culture, abolished the inhibitory influence of hypothalamic fragments upon prolactin secretion by the pituitary. Nevertheless, *in vivo* no direct evidence is available concerning the mode of action of perphenazine. BEN-DAVID (1968 c), therefore compared the mammotropic effect of perphenazine in intact and hypophysectomized adult rats, some of which also received replacement therapy, in order to clarify the role of pituitary and peripheral hormones in perphenazine-induced lactation.

Experiments were carried out without any estradiol priming and started 7 days after hypophysectomy. Priming with estrogen was avoided in this study, because estrogen may induce mammary duct growth and some end-bud growth (LYONS, 1958) and stimulate prolactin secretion from the pituitary (RAMIREZ and MCCANN, 1964). Animals received s.c. daily injections of perphenazine for 5 days, and they were sacrificed ony day after the last injection, and their mammotropic index determined.

Fig. 21 shows that increased doses of perphenazine treatment resulted in higher development of the mammary glands in the intact rats. A daily dose of 5 mg/kg of perphenazine for 5 days was able to change undeveloped glands into the well developed and secretory type. Moreover, the more the lobulo-alveolar system was developed, the less fat tissue was found in the perphenazine-treated intact rats. However, the same doses of perphenazine, when given to hypophysectomized animals, failed to induce any mammotropic effect even at the high dose of 10 mg/kg. Results of the mammotropic effect of 5 days' treatment with perphenazine (5 mg/kg/d), estradiol (0.01 mg/kg/d), hydrocortisone (10 mg/kg/d) and ovine prolactin (1 mg = 24 IU/kg/d), individually and combined, showed that perphenazine treatment alone in intact rats resulted in a MTI of 4.7, while such development was never achieved in hypophysectomized animals with any combination of treatment, comprising perphenazine and estradiol or hydrocortisone or ovine prolactin. In intact rats, the additional hormonal treatment improved the lactogenic effect achieved by perphenazine, but no initiation of milk secretion was found in any of the hypophysectomized rats receiving the same combination of treatment.

The failure of perphenazine to induce mammary gland development in the hypophysectomized animals at any dose level clearly

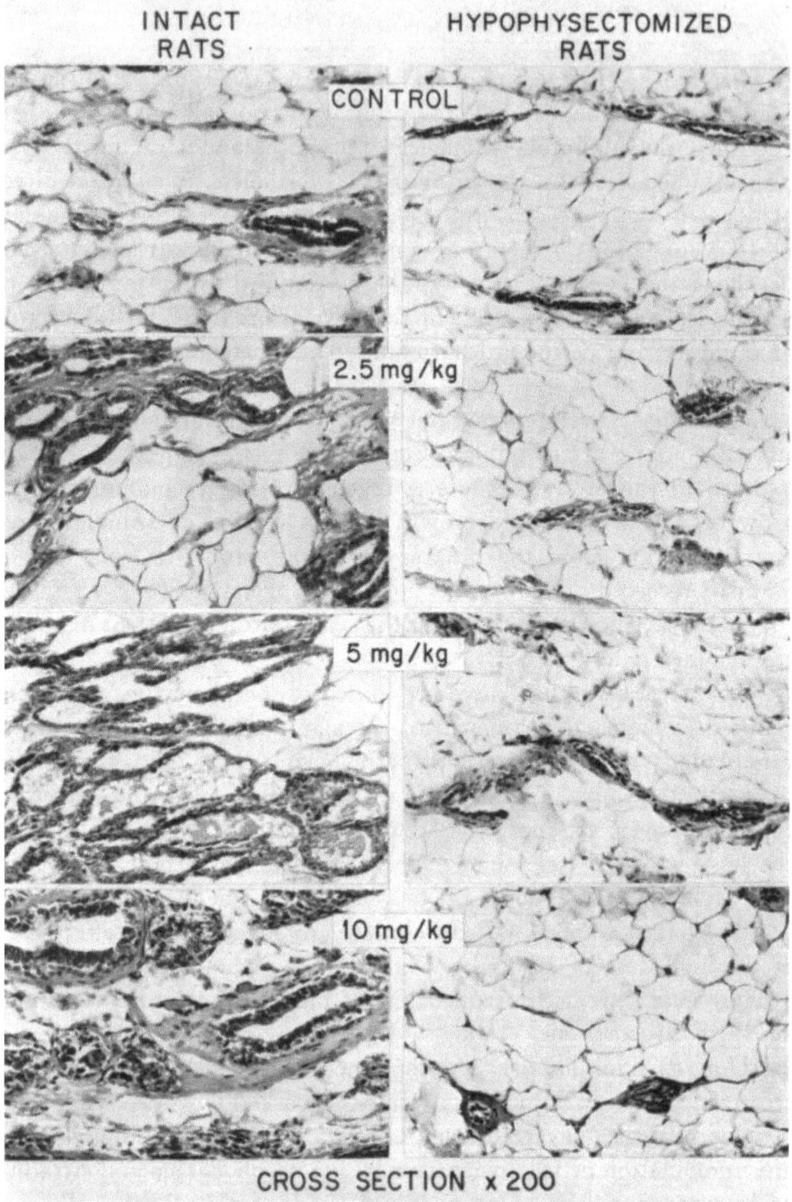

Fig. 21. Abolition of mammotropic effect of perphenazine after hypophysectomy: l: mammotropic reaction of intact rats after 5 days' treatment with 2.5, 5 and 10 mg/kg perphenazine respectively: MTI 5; r: complete absence of mammotropic reaction in hypophysectomized rats: MTI 1.5 after same treatment

indicates that the presence of the hypophysis is essential for this process, and that a drug endowed with mammotropic effect acts by stimulating secretion of mammotropic hormones from the pituitary.

Further evidence has been accumulated to show that perphenazine acts mainly via the hypothalamus, through stimulation of pituitary prolactin rather than other pituitary tropins. Moreover, the aforementioned results show that, in hypophysectomized rats, additional "replacement" treatment with steroid hormones was unable to produce a mammotropic effect similar to that achieved in intact animals by perphenazine alone. It is interesting that addition of ovine prolactin to the treatment failed substantially to improve the mammotropic effect in the hypophysectomized animals, a fact which could be explained by species differences in the source of prolactin (LEVY and SAMPLINER, 1962) or by the use of a subthreshold dose (24 IU/kg/d for 5 days), or by both together.

f) Conclusions

Drugs which depress hypothalamic activity, such as derivatives of phenothiazine, reserpine, butyrophenone, etc., can induce lactation. This effect cannot be prevented by hypothalamic energizers. The effect is specific and cannot be obtained by tranquilizing or sedative drugs which are not active at the hypothalamic level. Such drugs may evoke lesser degrees of lactation in as far as they affect the hypothalamus along with the cortical or internuncial synapses. In this case, over-priming with estrogen can unmask and potentiate their lactogenic effect. The sedative effect *per se* is not decisive for any lactogenic activity, indeed a separation of both properties seems possible. The mechanism of the lactogenic effect of all hypothalamic tranquilizers is based on the depression of the hypothalamic PIF; it depends on the presence of the hypothalamus and the pituitary.

Chapter 14

Mammotropic Effect of Phenothiazine Derivatives

a) History

Phenothiazine derivatives have a broad spectrum of pharmacological effects: antiemetic, analgetic, antihistaminic, adrenergic blocking, anticholinergic, antipsychotic, antipruritic, antitussive, anti-nauseant, antispasmotic, tranquilizing and mammogenic (HOLLISTER, 1964).

Galactorrhea in female patients, and breast development in males, after chlorpromazine treatment, were reported by various authors (WINNIK and TENNENBAUM, 1955; POLISHUK and KULCSAR, 1956; SULMAN and WINNIK, 1956; WINNIK and SULMAN, 1956; ROBINSON, 1957; HOOPER, WELCH, POINT and SHACKELFORD, 1961). Most authors related these effects to an interference with the normally-functioning endocrine release system of the hypothalamus. This view is strengthened today by a more exact knowledge of the mechanism of lactation, according to which depression of the hypothalamus causes suppression of the hypothalamic prolactin-inhibiting factor (PIF), thus enhancing prolactin secretion from the anterior pituitary.

Preliminary reports on the structure-activity relationship (SAR) of some phenothiazines with regard to their mammotropic effect were published by our Department (KHAZAN et al., 1962; BEN-DAVID, DIKSTEIN and SULMAN, 1965; KHAZAN, BEN-DAVID, MISHKINSKY, KHAZEN and SULMAN, 1966; KHAZEN, MISHKINSKY, BEN-DAVID and SULMAN, 1968 a, b) and by others (MEITES, 1957 a; BHARGAVA and CHANDRA, 1964; WELSH, 1964; ROSSI, 1964; LAUNCHBURY, 1964), mostly as secondary parameters of chemical and pharmacological research.

b) Mammotropic Index (MTI)

We undertook to screen all available phenothiazine and related structures in clinical and research use (tranquilizers and non-tranquilizers), in order to establish the rules of the SAR of their mammotropic effect. The two parameters used were the index of mammary development, "mammotropic index" (MTI), and body temperature drop (BTD) after subcutaneous injection of the substances. Measurement of these two parameters was performed as described in Chapter 26 — b.

Female albino rats of the Hebrew University "Sabra" strain, weighing 200 ± 10 g, were primed for 10 days s.c. with 8 μg/rat/day estradiol. On the 11th—15th day the drug to be tested was injected s.c. in doses of 5—40 mg/kg/day. On the 16th day the animals (5—10 per drug) were sacrificed and the MTI determined. Maximum development of the mammary glands of the controls (estradiol-primed for 10 days and saline-injected for the following 5 days) was found to reach MTI 2, with a mean MTI of 1.2. It is essential to keep strictly to a low priming dose (in our rats 8 μg/rat/day of estradiol) and to use female rats of 200 ± 10 g, because animals of larger weight or primed with bigger doses may tend to give a higher MTI and thus yield false positive results.

c) Body Temperature Drop (BTD)

Groups of 10 female rats (200 ± 10 g) were injected s.c. with a single dose of the phenothiazines. Temperatures were recorded at 0, 1, 2, 3, 4 and 5 hours with an electrocouple thermometer (Electric Universal Thermometer TE_3, "Ellab Instruments", Copenhagen). Control animals were injected with 0.2 ml saline s.c.

d) Screening of Phenothiazine Derivatives

Table 1 compares the MTI with the BTD effected by 87 phenothiazine derivatives at a dosage level of 5 and 40 mg/kg. It emerges that the most mammotropic phenothiazine derivatives were piperacetazine, perphenazine, propericiazine, thiopropazate, pipamazine, TPO-33, SKF 7100-A, 2-bromopromazine, fluphenazine, acetophenazine and SKF 7221-I.

It seems that a straight 3-C side chain originating at N_{10} is of primary importance because it may structurally form half a hexane ring, and that a heterocyclic ring at its end considerably increases mammotropic activity (cf. Chapter 19).

e) Screening of Phenothiazine Sulfoxides and Sulfodioxides

Our attention was drawn to the sulfoxide derivatives for three main reasons:

1. Only very mild oxidation is needed to convert phenothiazine into its lactogenic sulfoxide derivative, while more drastic oxidation yields the inactive sulfone derivative (Forrest and Piette, 1964).

2. The reversibility of the phenothiazine sulfoxide oxidation is very high: acetylpromazine and RP-8909 are oxidized by air and produce their sulfoxides very easily, while the respective sulfoxides are reduced to the parent substances only by exposure to light (Yamamoto and Fujisawa, 1963).

3. Mammalian drug metabolism will normally stop at the sulfoxide level (SO). Sulfones, which contain SO_2, are, therefore, not found in biological material like blood, urine and tissue.

Table 1. *Mammotropic index of 87 phenothiazine derivatives in oestrogen-primed female rats (200 ± 10 gm). Oestradiol priming: 8 μg/rat/day, for 10 days; phenothiazine treatment: 5—40 mg/kg/day for 5 days. Preparations are listed in descending order of their lactogenic effect at 5 mg/kg*

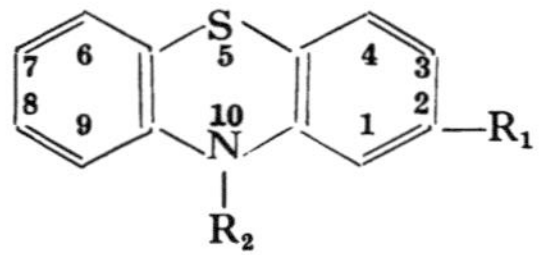

Generic and proprietary name	R_1	R_2	Daily dose mg/kg	Mammotropic index	Body temperature drop °C
Piperacetazine (Quide, Pitman-Moore)	$-C(=O)-CH_3$	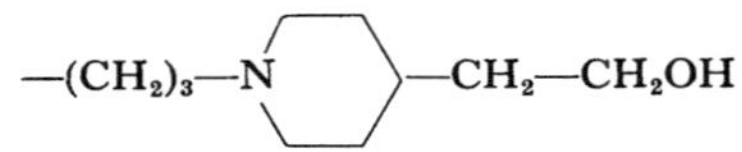	5 40	4.8 5.2	1.8 3.9
Perphenazine (Trilafon, Schering)	—Cl	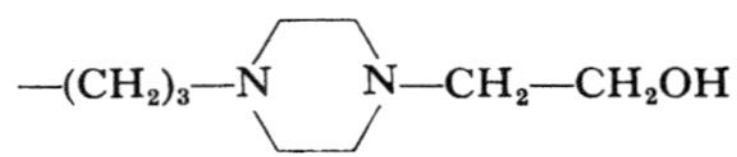	5 40	4.7 5.9	3.3 3.3
Propericiazine (Neuleptil, Specia) (SKF 20716)	—CN	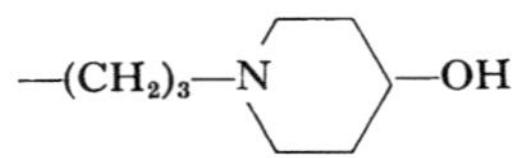	5 40	4.6 5.2	2.3 2.4
Thiopropazate (Dartal, Searle)	—Cl	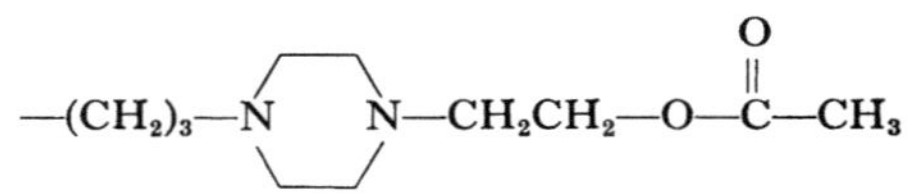	5 40	4.6 5.0	1.1 2.0

Pipamazine (Mornidine, Searle)	$-Cl$	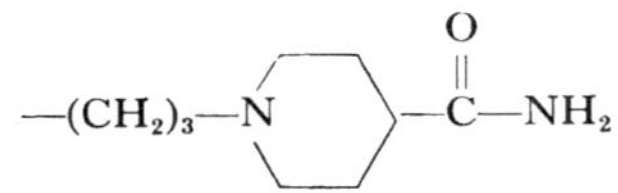	5 40	4.5 4.4	1.1 2.4
TPO-33 (Sandoz)	$-S(=O)-CH_3$	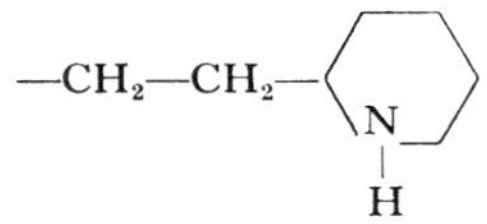	5 40	4.4 3.0	0 0.3
SKF 7100-A (Smith, Kline & French)	$-CF_3$	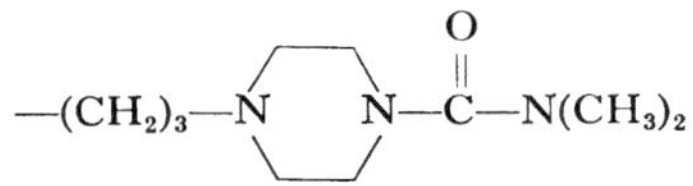	5 40	4.4 4.5	0 1.8
2-Bromopromazine (NIMH)	$-Br$	$-(CH_2)_3-N(CH_3)_2$	5 40	4.4 —	1.7 7.4
Fluphenazine (Prolixin, Squibb Permitil, White)	$-CF_3$	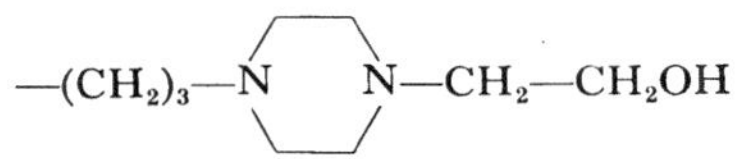	5 40	4.2 5.6	0 3.4
Acetophenazine (Tindal, Schering)	$-C(=O)-CH_3$	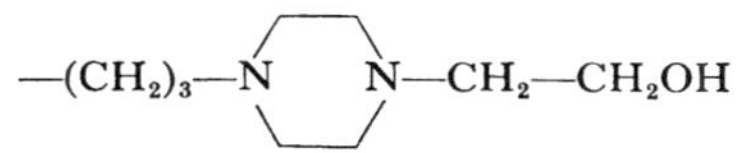	5 40	4.2 3.5	0.9 2.1
SKF 7221-I (Smith, Kline & French)	$-CF_3$	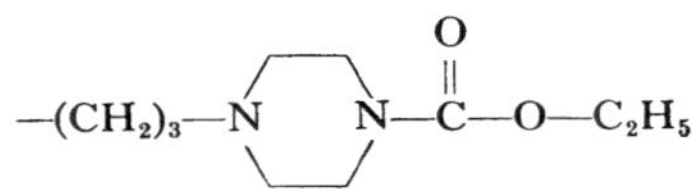	5 40	3.8 3.8	1.1 2.0

Table 1 (continued)

Generic and proprietary name	R_1	R_2	Daily dose mg/kg	Mammotropic index	Body temperature drop °C
SKF 7136-A (Smith, Kline & French)	$—CF_3$	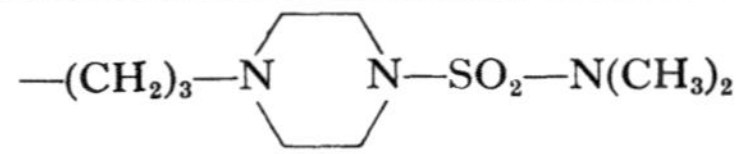	5 40	3.8 4.2	1.3 0.8
Carphenazine (Proketazine Wyeth)	$—C(=O)—C_2H_5$	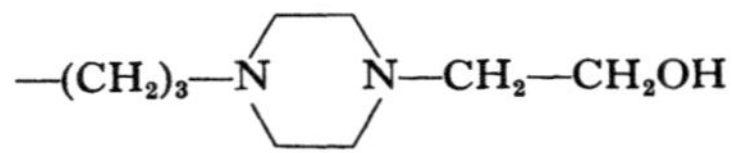	5 40	3.8 3.4	0.6 1.6
Pasaden (Chemiewerk Homburg)	$—CH_3$	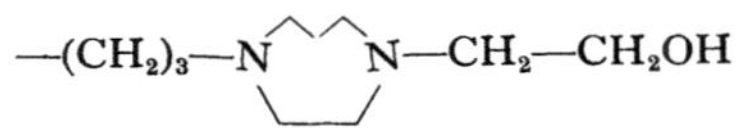	5 40	3.8 1.9	0 1.1
Ethylbutrazine (Sergetyl, Vaillant) (Ethymemazine, Specia) (RP 6484)	$—C_2H_5$	$—CH_2—CH(CH_3)—CH_2—N(CH_3)_2$	5 40	3.6 4.2	0.3 2.2
SKF 6063-A (Smith, Kline & French)	$—Cl$	$—CH_2—CH_2—C(=O)—N(H)—(CH_2)_3—N(CH_3)_2$	5 40	3.6 3.1	0.4 0.4
Butyrylperazine (Randolectil, Bayer)	$—C(=O)—C_3H_7$	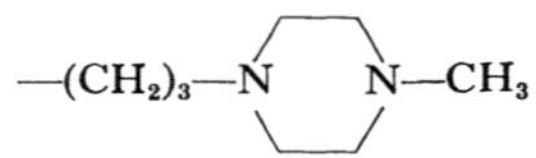	5 40	3.5 5.2	1.1 4.8

Thiethylperazine (Torecan, Sandoz)	$-SC_2H_5$	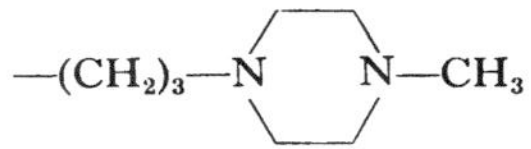	5 40	3.4 3.9	0 0
Acepromazine (Plegicil, Byla)	$-C(=O)-CH_3$	$-(CH_2)_3-N(CH_3)_2$	5 40	3.2 4.3	1.6 4.2
Trifluorperazine (Stelazine, SKF)	$-CF_3$	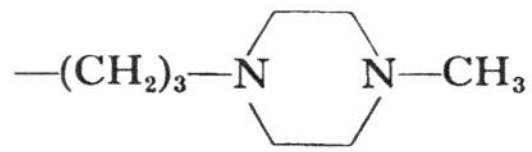	5 40	3.2 5.4	0.1 0.4
Thioridazine (Mellaril, Sandoz)	$-S-CH_3$	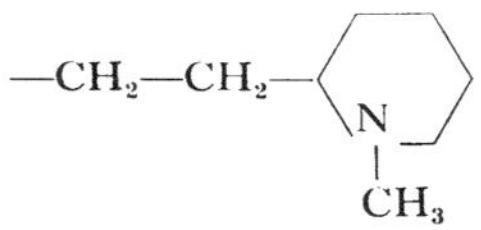	5 40	3.2 3.0	1.1 0.6
Methophenazine (Frenelon, Un. Works, Pharm. & Diet. Prod., Budapest)	$-Cl$	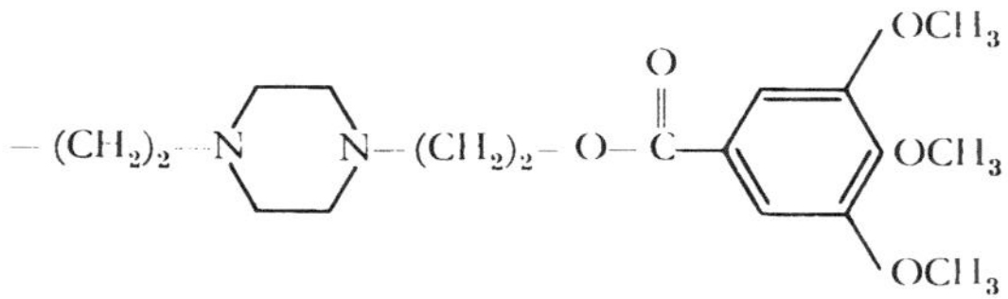	5 40	3.2 2.7	0.7 0
SKF 7887-A_2 (Smith, Kline & French)	$-Cl$	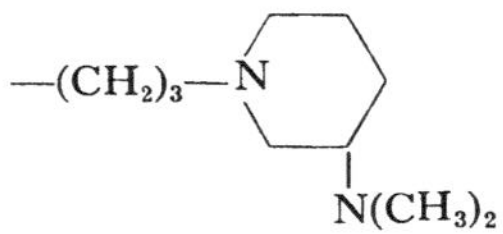	5 40	3.2 1.8	0 0.9
SKF 8448-I (Smith, Kline & French)	$-Cl$	$-(CH_2)_3-NH-CH_2-CF_3$	5 40	3.2 2.7	0 0.6

Table 1 (continued)

Generic and proprietary name	R_1	R_2	Daily dose mg/kg	Mammotropic index	Body temperature drop °C
SKF 5910-I_2 (Smith, Kline & French)	$-Cl$	$-CH_2-CH(CH_2-N(CH_3)_2)-CH_2-N(CH_3)_2$	5 40	3.2 3.4	0 0.4
UCB 2493 (Chloroesucos UCB, Brussels)	$-Cl$	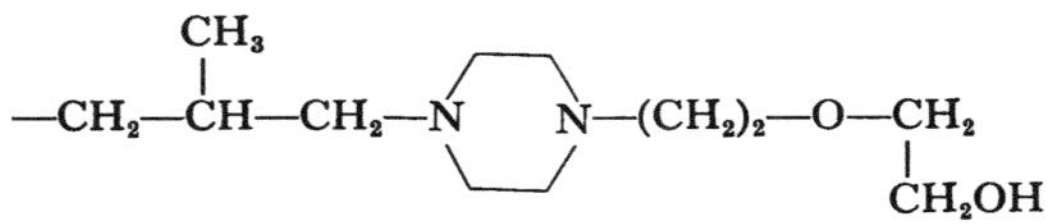	5 40	3.2 4.4	0.3 1.8
Cyamepromazine (R. P. 7204, Specia)	$-CN$	$-CH_2-CH(CH_3)-CH_2-N(CH_3)_2$	5 40	3.2 5.2	3.0 6.0
TPN 12 (Sandoz)	$-SO_2-CH_3$	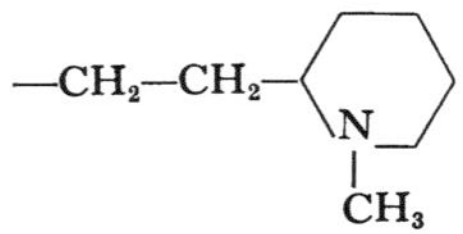	5 40	3.0 4.0	1.3 2.5
Chlorpromazine (Largactil, Specia Thorazine, SKF)	$-Cl$	$-(CH_2)_3-N(CH_3)_2$	5 40	3.0 3.6	2.8 9.4

SKF 6058-I_2 (Smith, Kline & French)	$-Cl$	$-(CH_2)_3-N(CH_3)-(CH_2)_3-N(CH_3)_2$	5 40	3.0 3.1	0 1.6
2-Chlorophenothiazine (SKF 3227)	$-Cl$	$-H$	5 40	3.0 2.5	0 0.8
Prochlorperazine (Compazine, SKF) (R.P. 6140 Specia)	$-Cl$	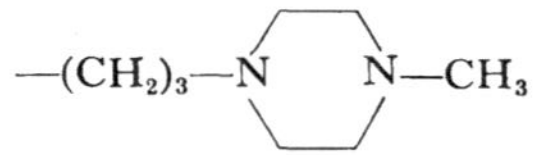	5 40	3.0 4.4	1.4 0.6
Chlorproethazine (R.P. 4909, Specia)	$-Cl$	$-(CH_2)_3-N(C_2H_5)_2$	5 40	2.8 3.4	0 2.5
Perazine (Taxilan, Promonta)	$-H$	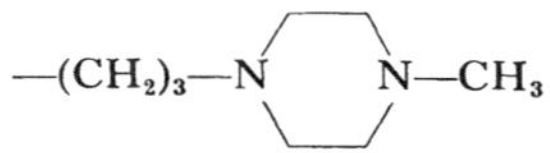	5 40	2.8 3.5	0 0.1
SKF 6064-J (Smith, Kline & French)	$-Cl$	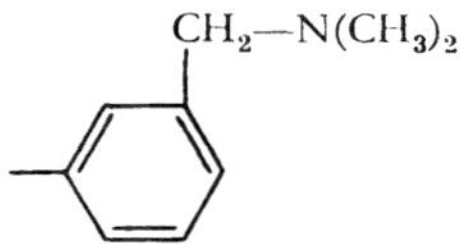	5 40	2.8 2.4	1.2 1.2
SKF 6108-J_2 (Smith, Kline & French)	$-CF_3$	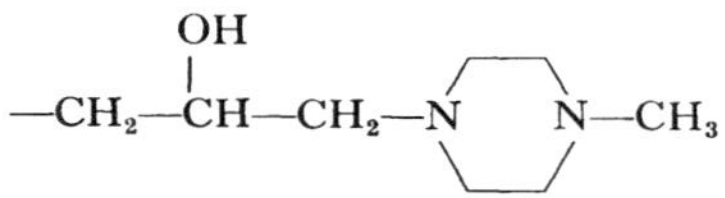	5 40	2.8 3.0	0 0.2

Table 1 (continued)

Generic and proprietary name	R_1	R_2	Daily dose mg/kg	Mammotropic index	Body temperature drop °C
SKF 8745 (Smith, Kline & French)	$-Cl$	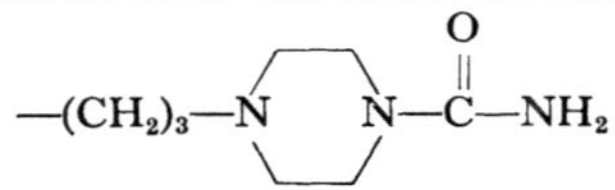	5 40	2.8 3.5	2.3 1.9
Thioperazine (Majeptil, Specia) (Vontil, SKF) (RP 7843)	$-SO_2N(CH_3)_2$	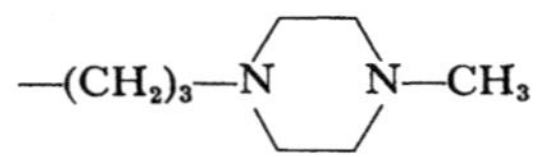	5 40	2.6 2.4	0 0
Alimemazine R.P. 6549, Specia (Temaril, SKF)	$-H$	$-CH_2-CH(CH_3)-CH_2-N(CH_3)_2$	5 40	2.6 2.9	0.9 1.6
SKF 5757-G (Smith, Kline & French)	$-Cl$	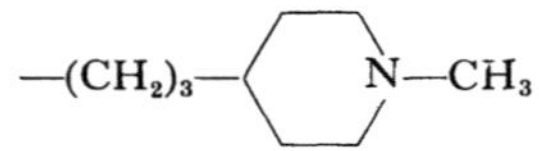	5 40	2.6 4.4	0 0
SKF 7009-I (Smith, Kline & French)	$-CF_3$	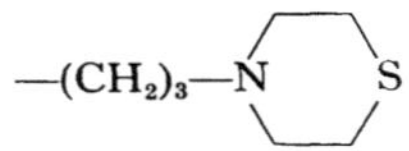	5 40	2.6 3.7	0.2 0.5
SKF 6051-A_2 (Smith, Kline & French)	$-Cl$	$-(CH_2)_3-NH(CH_2)_3-N(CH_3)_2$	5 40	2.6 4.3	0.6 2.2

R.P. 6710 (Specia)	—Cl	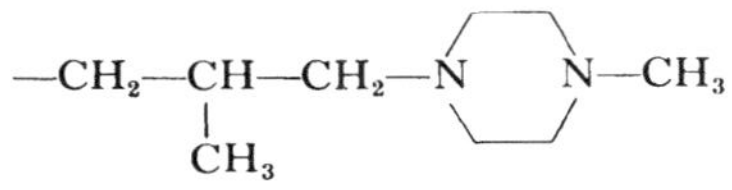	5 40	2.5 3.8	0.8 1.4
7-OH-Chlorpromazine (NIMH)	—Cl	$-(CH_2)_3-N(CH_3)_2$ and 7-OH	5 40	2.5 3.6	— —
3-Hydroxypromazine (NIMH)	3-OH	$-(CH_2)_3-N(CH_3)_2$	5 40	2.5 4.3	0.7 2.6
8-Hydroxychlorpromazine (NIMH)	—Cl 8-OH	$-(CH_2)_3-N(CH_3)_2$	5 40	2.5 2.6	0 0.3
2-Methoxyphenothiazine (NIMH)	$-OCH_3$	—H	5 40	2.5 3.6	1.0 1.1
Diethazine (Diparcol, Specia)	—H	$-(CH_2)_2-N(C_2H_5)_2$	5 40	2.4 2.9	0 0
Dipyripromazine (Promoton, Galma)	—Cl and 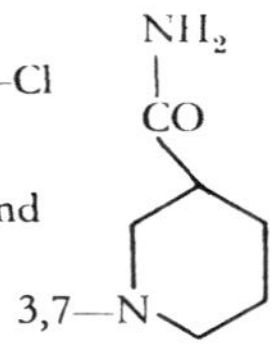	$-(CH_2)_3-N(CH_3)_2$	5 40	2.4 2.8	2.0 5.0

Table 1 (continued)

Generic and proprietary name	R_1	R_2	Daily dose mg/kg	Mammotropic index	Body temperature drop °C
Mepazine (Pacatal, Promonta)	—H	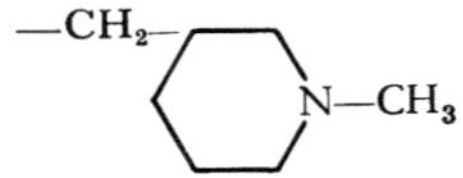	5 40	2.4 3.2	0 3.2
SKF 6693-A_2 (Smith, Kline & French)	—Cl	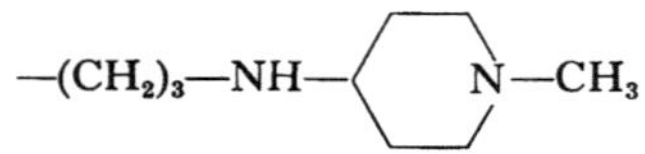	5 40	2.4 3.2	0 0.5
SKF 5611-A (Smith, Kline & French)	—Cl	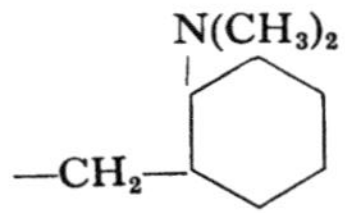 Cis isomer	5 40	2.4 2.5	0.7 0.3
SKF 3072-A (Smith, Kline & French)	4-Cl	$—(CH_2)_3—N(CH_3)_2$	5 40	2.4 3.5	0.7 1.1
SKF 7838 (Smith, Kline & French)	—Cl	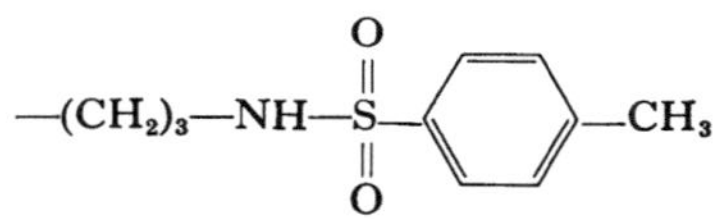	5 40	2.4 3.4	0.3 1.4

Mesoridazine (NC-123, Sandoz)	$-S(=O)-CH_3$	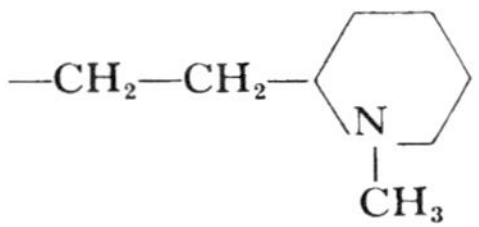	5 40	2.2 3.9	0 0.8
SKF 6262-A_2 (Smith, Kline & French)	—Cl	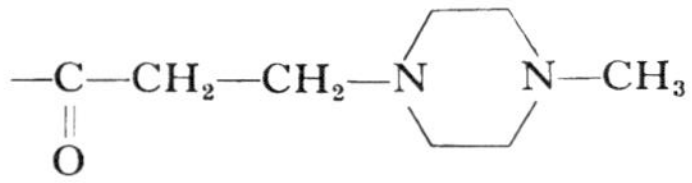	5 40	2.2 2.8	0 0
Methoridazine (KS-33, Sandoz)	$-OCH_3$	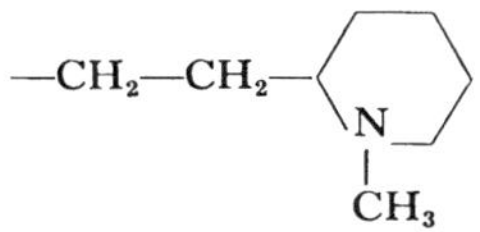	5 40	2.2 3.6	0.4 1.0
Promazine (Sparine, Wyeth)	—H	$-(CH_2)_3-\overset{+}{N}(CH_3)_2$	5 40	2.6 2.2	0.8 3.4
Compound 22 (Winthrop)	—H	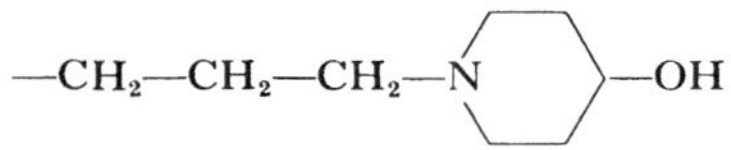	5 40	2.2 3.6	1.3 3.1
SKF 2680-J (Smith, Kline & French)	—Cl	$-(CH_2)_3-\overset{+}{N}(CH_3)_3$	5 40	2.2 3.3	1.0 1.1
SKF 12116-A (Smith, Kline & French)	—Cl	$-C(=O)-CH_2-CH_2-N(C_2H_5)_2$	5 40	2.0 3.0	0 0.9

Table 1 (continued)

Generic and proprietary name	R_1	R_2		Daily dose mg/kg	Mammotropic index	Body temperature drop °C
SKF 5612-A (Smith, Kline & French)	$-Cl$	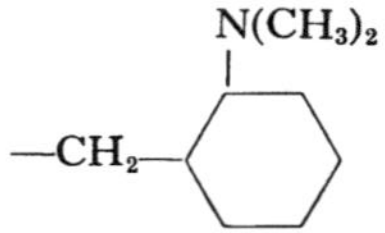	Trans isomer	5 40	2.0 3.3	1.5 1.1
KS-75 (Sandoz)	$-OC(H)(CH_3)_2$	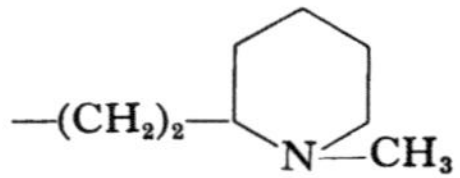		5 40	2.0 2.6	0.9 1.6
Methylene blue	$-H$ 3, 7-$N(CH_3)_2$	$-H$		5 40	1.9 2.9	1.3 3.5
Profenamine Parsidol, R.P. 3326 (Specia)	$-H$	$-CH_2-CH(CH_3)-N(C_2H_5)_2$		5 40	1.9 2.0	0.1 3.0
Methdilazine (Tacaryl, Mead-Johnson)	$-H$	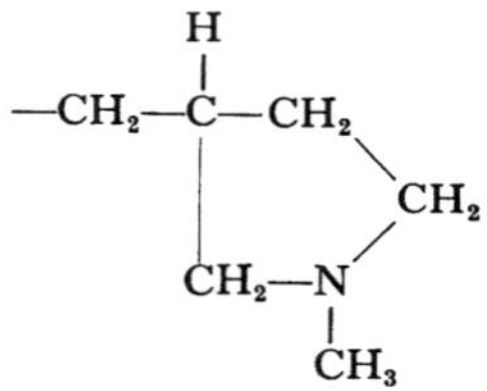		5 40	1.8 2.6	2.5 2.7

Methopromazine (Mopazine, Specia)	$-OCH_3$	$-(CH_2)_3-N(CH_3)_2$	5 40	1.8 1.4	1.0 3.0
Propionylpromazine (Combelen, Bayer)	$-C(=O)-C_2H_5$	$-(CH_2)_3-N(CH_3)_2$	5 40	1.8 3.6	1.0 2.0
SKF 13813-A (Smith, Kline & French)	$-C(=O)-CH_3$	$-C(=O)-CH_2-CH_2-N(CH_3)_2$	5 40	1.8 2.6	0.9 1.1
SKF 4523 (Smith, Kline & French)	$-H$	$-C(=S)-N(C_2H_5)_2$	5 40	1.8 3.3	0.2 1.4
Dixyrazine (Esucos UCB, Brussels)	$-H$	$-CH_2-CH(CH_3)-CH_2-N\langle\,\rangle N-(CH_2)_2-O(CH_2)_2OH$	5 40	1.7 4.2	0.6 2.2
3-Chlorophenothiazine (SKF 18722)	3-Cl	$-H$	5 40	1.6 2.2	0.1 1.0
Levomepromazine (Nozinan, Specia)	$-OCH_3$	$-CH_2-CH(CH_3)-CH_2-N(CH_3)_2$	5 40	1.6 1.6	1.9 4.6

Table 1 (continued)

Generic and proprietary name	R_1	R_2	Daily dose mg/kg	Mammotropic index	Body temperature drop °C
Dimethoxanate (Clothera, Ayerst)	$-H$	$-C(=O)-O-(CH_2)_2-O-(CH_2)_2-N(CH_3)_2$	5 40	1.6 1.8	0 1.6
Phenothiazinyl-propionitril	$-H$	$-CH_2-CH_2-CN$	5 40	1.6 1.8	0.3 1.4
Dimetiotazine (R.P. 8599 Specia)	$-SO_2-N(CH_3)_2$	$-CH_2-CH(CH_3)-N(CH_3)_2$	5 40	1.6 2.5	0.6 1.0
SKF 6045-A_2 (Smith, Kline & French)	$-Cl$	$-(CH_2)_3-N(CH_3)-(CH_2)_2-N(CH_3)_2$	5 40	1.6 3.2	0.5 0.9
10-phenothiazinyl-propionic acid (self synthesis)	$-H$	$-(CH_2)_2COOH$	5 40	1.6 2.2	0 0
Aminopromazine (Lispamol, Specia)	$-H$	$-CH_2-CH(N(CH_3)_2)-CH_2-N(CH_3)_2$	5 40	1.5 2.5	0 0.3

Promethazine (Phenergan, Specia)	—H	$-CH_2-CH(CH_3)-N(CH_3)_2$	5 40	1.4 2.3	0.7 2.1
Propriomazine (Largon, Wyeth) (Dorevane, Clin)	$-C(=O)-CH_2CH_3$	$-CH_2-CH(CH_3)-N(CH_3)_2$	5 40	1.4 3.2	0.2 1.0
Pyrathiazine (Pyrrolazote, Upjohn)	—H	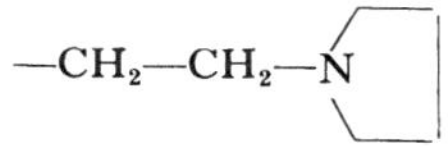	5 40	1.4 2.4	0 1.0
6-OH-Chlorpromazine (NIMH)	—Cl	$-(CH_2)_3-N(CH_3)_2$ and 6-OH	5 40	1.4 2.1	— —
1-Chloro-phenothiazine (SKF 24200)	1-Cl	—H	5 40	1.4 2.2	0 0.3
Lauth Violet (Thionin, Merck)	3-amino, 7-imino-	—H	5 40	1.4 2.6	0.4 1.0
Toluidine Blue Tolonium (Blutene, Abbott)	3-$N(CH_3)_2$ 5-Cl 7-NH_2 8-CH_3	—H	5 40	1.4 2.3	1.0 2.0
Phenothiazine (Specia)	—H	—H	5 40	1.2 1.8	0.2 0.6
Saline (control)	—	—	0.5 ml	1.2	0

Table 2. *Mammotropic index of 15 phenothiazine sulfoxides and sulfodioxides in oestrogen-primed female rats (200 ± 10 gm). Oestradiol priming: 8 μg/rat/day for 10 days; treatment: 5—200 mg/kg/day, for 5 days. Preparations are listed in descending order of their lactogenic effect at 5 mg/kg*

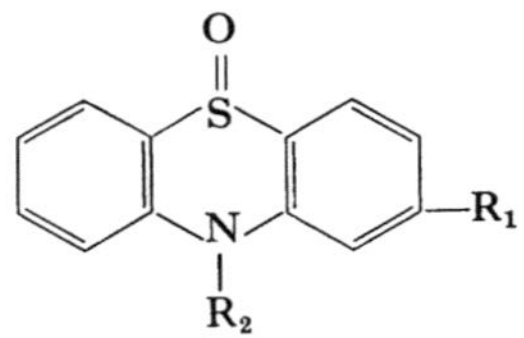

Generic and proprietary name	R_1	R_2	Daily dose mg/kg	Mammotropic index	Body temperature drop °C
Perphenazine sulfoxide (Schering, USA)	—Cl	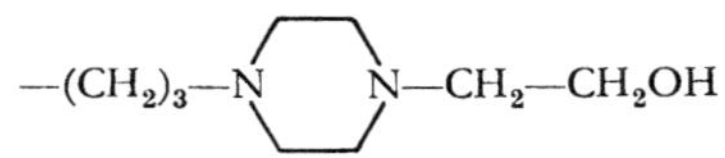	5 40 200	3.8 4.3 5.0	0 0 0.7
Piperacetazine sulfoxide (self synthesis)	$—C(=O)—CH_3$	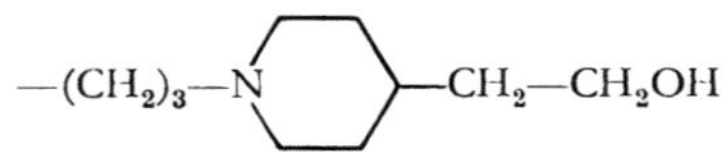	15 40 200	3.6 4.4 5.0	— —
R.P. 8.580 (Specia)	$—SO_2—N(CH_3)_2$	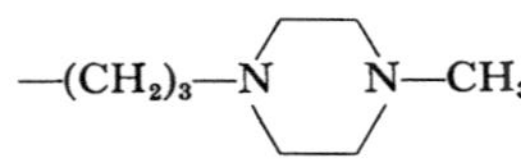	5 40 200	3.3 4.9 5.0	0 0.3 0.7

Methdilazine sulfoxide (Mead Johnson)	—H	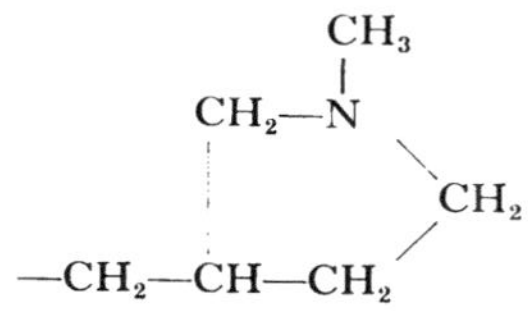	5 40 200	2.6 3.0 4.4	0 1.0 1.3
SKF 4611-A sulfodioxide (Smith, Kline & French)	—Cl	$(CH_2)_3$—$N(CH_3)_2$	5 40 200	2.6 3.3 3.9	0 0.1 1.2
Promethazine sulfoxide (Phenergan sulfoxide, Specia)	—H	$—CH_2—CH(CH_3)—N(CH_3)_2$	5 40 200	2.5 2.8 2.8	0.8 2.1 1.9
SKF 5419-A Triflupromazine sulfoxide, (Smith, Kline & French)	—CF_3	—$(CH_2)_3$—$N(CH_3)_2$	5 40 200	2.4 3.2 4.7	0 1.4 2.7
SKF 17910-A (Smith, Kline & French)	—Cl	$—C(=O)—(CH_2)_2—N(CH_3)_2$	5 40 200	2.2 2.5 4.0	— — —
SCH 13361 (Schering)	—Br	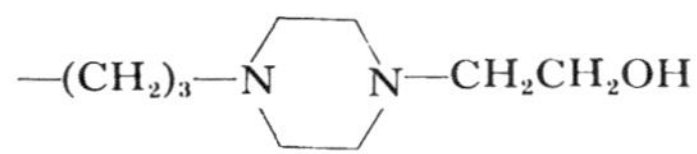	5 40 200	2.2 2.6 3.8	— — —

Table 2 (continued)

Generic and proprietary name	R_1	R_2	Daily dose mg/kg	Mammotropic index	Body temperature drop °C
SKF 7145-A sulfoxide (Smith, Kline & French)	—Cl	$-(CH_2)_3-N(H)-CH_3$	5	1.8	0
			40	2.5	0.2
			200	3.0	0.5
Chlorpromazine sulfoxide (R.P. 5293, Specia) (SKF 4260-A)	—Cl	$-(CH_2)_3-N(CH_3)_2$	5	1.9	0.5
			40	2.8	0
			200	4.0	0.5
Promazine sulfoxide (Wy 1193, Wyeth)	—H	$-(CH_2)_3-N(CH_3)_2$	5	1.8	0
			40	2.8	0
			200	3.4	2.8
Phenothiazine sulfoxide	—H	—H	5	1.6	0.9
			40	2.2	0
			200	3.0	1.2
Oxomemazine sulfodioxide (Doxergan, Specia)	—H	$-CH_2-CH(CH_3)-CH_2-N(CH_3)_2$	5	1.4	0.8
			40	2.6	1.0
			200	4.0	1.9
SKF 5332-I sulfoxide (Smith, Kline & French)	$-OCH_3$	$-(CH_2)_3-N(CH_3)_2$	5	1.4	—
			40	—	—
			200	—	—

Testing of 15 sulfoxide derivatives of phenothiazine revealed that oxidation decreased considerably the sedative and tranquilizing effect as measured by body temperature drop. The mammotropic activity, however, was much less affected. This opened a way to synthesis of lactogenic drugs practically free of tranquilizing side effects. The results are presented in Table 2. The most mammotropic phenothiazine sulfoxides were perphenazine sulfoxide, RP-8580, and methdilazine sulfoxide.

In Tables 1 and 2 the phenothiazine derivatives studied are listed in descending order of their lactogenic effect at 5 mg/kg. This arrangement conspicuously shows that special substitutions at C_2 and N_{10} of the phenothiazine molecule are essential for eliciting strong mammotropic activity. Substitution of C_2 by Cl, Br, CF_3, SCH_3, SC_2H_5, CN or $CO(CH_3)$ increased the mammotropic effect. Likewise substitution at N_{10} by piperidine, pyrrolidine and piperazine rings provided highly mammotropic compounds (cf. Chapter 19). The preparations, which yielded a mammotropic index of 2.0 or less, may be regarded as completely ineffective compounds. Moreover, we showed that no parallelism exists between the mammotropic and tranquilizing effects of several phenothiazines, e. g. perphenazine sulfoxide and chlorpromazine sulfoxide. Both compounds showed only very slight tranquilizing activity or none at all, but they still proved to be strongly mammotropic. The tables present also the body temperature drop as a measure of tranquilizing activity.

Table 2 shows the lactogenic effect of 15 phenothiazine sulfoxides. Comparison of their lactogenic activity with that of their phenothiazine parent compounds (containing S instead of SO) shows a clearly diminished but still good lactogenic activity which can be enhanced to maximum effect by increased dosage. For practical purposes this would be no drawback, since sulfoxides have no tranquilizing effect unless their dosage is increased beyond a 20-fold concentration (Hotovy and Kapff-Walter, 1960).

Rossi (1964) pointed out that phenothiazine derivatives of the piperazine series generally possess a greater antipsychotic activity per milligram than those of the dimethylamine or piperidyl series. Within these three series, differences in psychotherapeutic activity are also attributable to the nature of the substitution at position 2 on the phenothiazine nucleus. As a rule, the chlorine-substituted derivatives are less active than the trifluoromethylsubstituted compounds. Mercaptomethyl substitution appears to imbue the compound with an activity approximately equal to that of the chlorine substitution. Phenothiazines, lacking a substitution in position 2, or having an acetyl group, appear to possess the lowest anti-psychotic potency within their respective series. Sulfoxide formation nearly abolishes the anti-psychotic effect, without greatly diminishing the lactogenic effect. It is interesting to note that most of the rules pertaining to the psychotherapeutic activity hold also—though not necessarily—for the lactogenic effect. We know now that the common denominator for both actions is a 3—4 ring system which, in its structural arrangement, is similar to that of the steroids and therefore hypothalamo-tropic and hypophysio-tropic (cf. Chapter 19).

f) Stimulation of Synthesis and Release of Prolactin by Perphenazine

As previously described, perphenazine (Trilafon)—an amino derivative of chlorphenothiazine—was found to be a potent lactogenic agent in intact virgin rats (Ben-David, Dikstein and Sulman, 1965) but entirely inactive in hypophysectomized animals (Ben-David, 1968 c). Perphenazine was also reported to stimulate luteotropic activity (Ben-David, 1968 b) and to stimulate prolactin secretion *in vitro* when added to pituitary-hypothalamus co-culture (Danon, Dikstein and Sulman, 1963). Hence it was postulated that perphenazine exerts its lactogenic effect through stimulation of pituitary prolactin secretion. This stimulation is thought to be effected by interference with the action of the prolactin-inhibiting factor (PIF) released from the hypothalamus (Pasteels, 1961 a; Grosvenor, McCann and Naller, 1965; Schally, Kuroshima, Ishida, Redding and Bowers, 1965; Ben-David, 1968 c). However, the processes taking place in the hypothalamus and pituitary subsequent to the administration of perphenazine, leading to the ultimate release and biosynthesis of prolactin, have not been elucidated.

Ben-David and Chrambach (1969), therefore, studied the simultaneous effect of daily injections of perphenazine (5 mg/kg/d) for 1—5 days on mammary differentiation, pituitary weight and on pituitary and prolactin levels in female rats, using a modification of the pigeon-crop DNA method (Ben-David, 1967). Blood prolactin was extracted as described in Chapter 26—d.

The development and histological patterns of the mammary glands of control animals (prior to treatment) and animals after varying numbers of daily injections of 5 mg perphenazine/kg/day are shown in Figs. 22 and 23. The histological patterns reveal remarkable mammary differentiation in direct proportion to the duration of perphenazine treatment. Initiation of lobulo-alveolar development occurred after 2 days of treatment, and mammary secretion appeared in the glands after 3 days of treatment. After 4 days of daily perphenazine administration, and, similarly, after 5 days of treatment, mammary secretion was abundant.

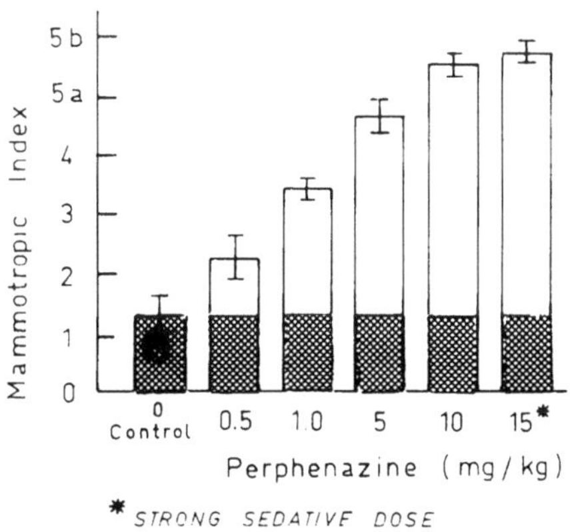

Fig. 22. Mammotropic effect (MTI) of increasing doses of perphenazine (0.5—15 mg/kg/day), given during 5 days to adult female rats. The MTI exceeds 3 after a daily dose of 1 mg perphenazine, 5 or 10 mg are optimum dosages, 15 mg have a strong sedative effect

The relative increase in the wet weight of pituitaries per 100 g body weight, following various periods of perphenazine treatment, is shown in Fig. 24. A significant increase of 32.4% ($P < 0.05$) was found 24 hours after a single injection of 5 mg perphenazine/kg, and a significant average increase of 40% ($P < 0.01$) was found after 2, 3, 4, and 5 days of treatment. Since the sedative action of the drug produced diminished food intake and loss of body weight in the test animals, a control series of non-treated rats, subjected to fasting for varying periods ranging between 1 and 5 days, was studied. In these animals there was a parallel decline in both pituitary and body weights, so that the quotient of pituitary weight divided by body weight was practically unchanged. Thus, the reported increases in pituitary weight per 100 g bodyweight represent not only increases in protein mass but,

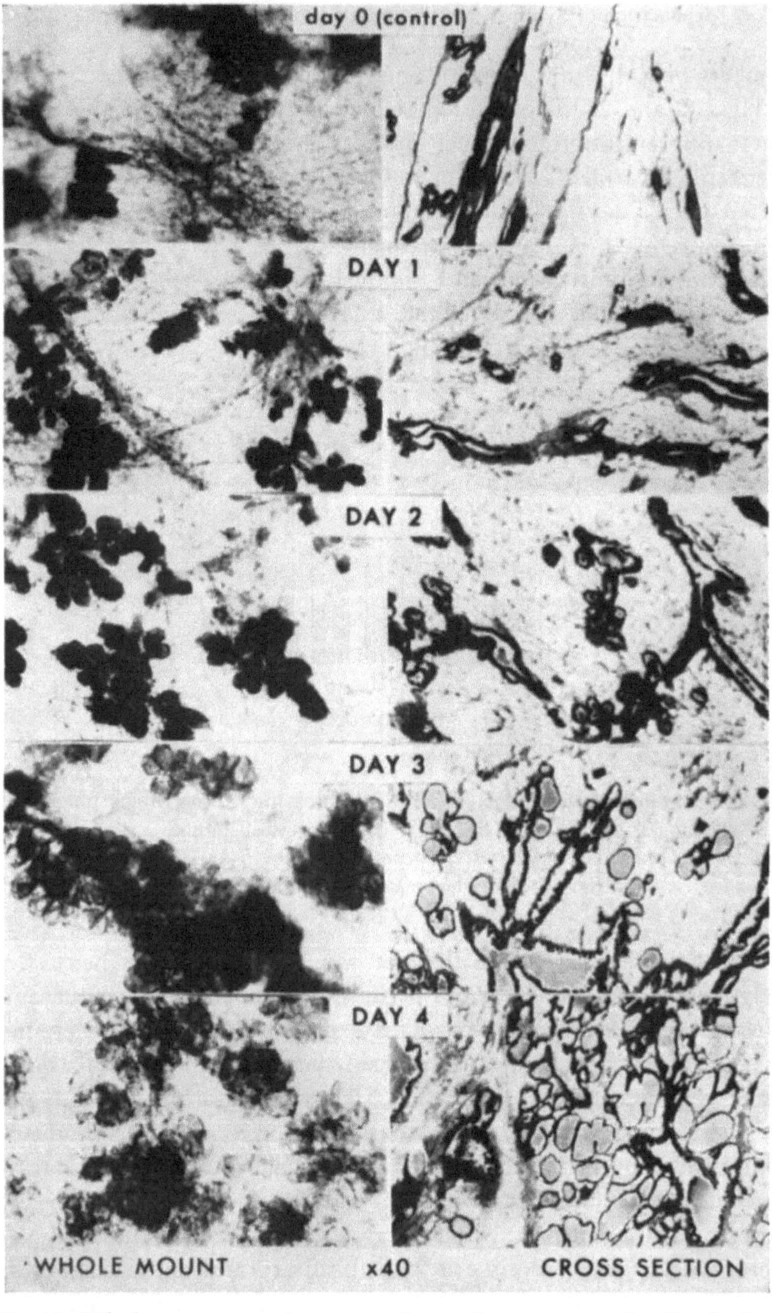

Fig. 23. Whole mounts and cross sections of mammary glands of virgin female rat which has received daily injections of 5 mg/kg perphenazine. Note the appearance of secretion starting after three days of treatment and the highly developed lobulo-alveolar system after four days of treatment

more specifically, the absence of any decline in the pituitary weight of perphenazine-treated animals together with the loss in total body weight.

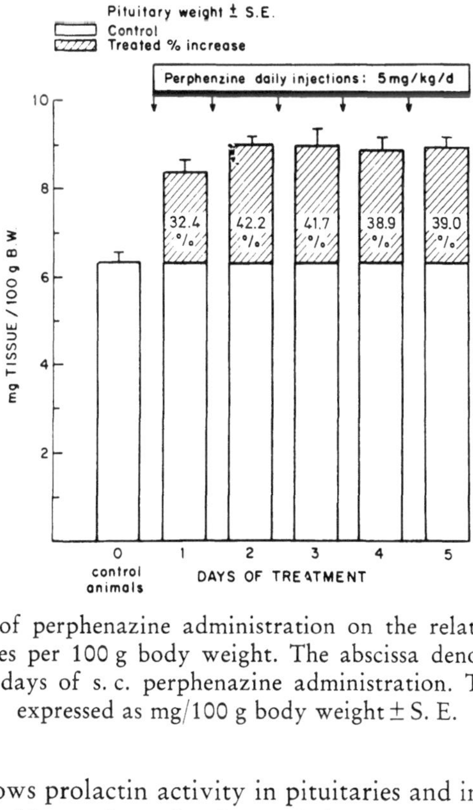

Fig. 24. Effect of perphenazine administration on the relative wet weight of rat pituitaries per 100 g body weight. The abscissa denotes the number of consecutive days of s. c. perphenazine administration. The weights are expressed as mg/100 g body weight ± S. E.

Fig. 25 shows prolactin activity in pituitaries and in serum of rats treated daily with perphenazine for periods up to 5 days. During this 5-day treatment, the initial release of prolactin from the pituitary into the blood and the subsequent production of prolactin in the pituitary can be distinguished kinetically. Twenty-four hours after a single injection of perphenazine, prolactin activity of the pituitary decreased to 48% ($P < 0.05$) of that of the untreated control animal. Concurrently, prolactin activity of the blood increased from 0.5 to 3.3 mU/ml ($P < 0.001$). By the 3rd day of treatment, pituitary prolactin per mg of tissue leveled off at 55% above the rate of the controls; the weight of the pituitary increased simultaneously with the relative amount of prolactin (activity per total weight of tissue).

This shows that the mammotropic action of perphenazine is mediated by pituitary prolactin, since the drug does not cause mammary growth in hypophysectomized animals (BEN-DAVID, 1968 c). The action of the drug on the pituitary gland is indirect; it is directed primarily at the hypothalamus, since implantation of perphenazine

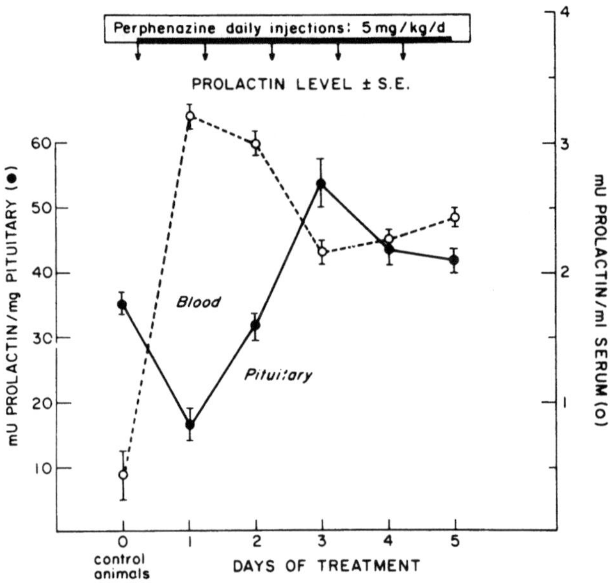

Fig. 25. Effect of perphenazine on pituitary and blood prolactin activity. The abscissa denotes the number of consecutive days of perphenazine administration. The vertical lines indicate the standard error. Note the immediate increase in blood prolactin heralded by its depletion from the pituitary

into the median eminence causes lactation in rabbits (MISHKINSKY, LAJTOS and SULMAN, 1966). Action of the drug on the pituitary gland is generally considered to be mediated by the hypothalamic prolactin-inhibiting factor (PIF) (PASTEELS, 1962; SCHALLY et al., 1965; GROSVENOR et al., 1965). Direct evidence that perphenazine stimulated prolactin secretion from the pituitary by suppressing the PIF is found in the work of DANON et al. (1963). It was shown that the addition of perphenazine to a pituitary-hypothalamus combined organ culture abolished the inhibitory effect of the hypothalamic fragments upon prolactin secretion by the pituitary. The blocking action of perphenazine on the PIF appears to be reversible and temporary, as diestrus in rats—a concomitant of its administration—terminates 7 days after treatment with the drug, with resumption of a normal estrus cycle (BEN-DAVID, 1968 b).

Perphenazine treatment seems to affect mainly pituitary prolactin rather than other hormones. This hypothesis is supported by the fact that perphenazine maintains full mammotropic activity in propylthiouracil-treated rats (BEN-DAVID, DIKSTEIN and SULMAN, 1966) and also in ovariectomized or adrenalectomized rats when administered immediately after the operation (BEN-DAVID, 1968 b, c; BEN-DAVID and SULMAN, 1969). Moreover, other phenothiazine derivatives, such as chlorpromazine, are reported to cause inhibition rather than stimulation of FSH and LH secretion (SULMAN and WINNIK, 1956; WINNIK and SULMAN, 1956; ALLOITEAU, 1957).

Whether perphenazine, mediated by PIF, regulates prolactin release or production in the pituitary remains to be established. The biphasic kinetics of simultaneous prolactin assays in the pituitary and in the blood depicted in Fig. 25 support the hypothesis that the action of the drug involves both release and production of prolactin. It remains to be shown to what extent the initial release-phase also comprises hormone production; and whether the production of the hormone involves biosynthesis or activation of a precursor protein.

The following facts suggest that perphenazine treatment results in "turning on" the pituitary: Following one day of treatment with the drug, pituitary weight significantly increases (Fig. 24), in spite of the pronounced depletion of prolactin (Fig. 25). Subsequent to the release of the hormone, its activity increases with time and reaches a steady state of secretion significantly above the level characteristic for the control gland (Figs. 24 and 25). It seems plausible, in view of the apparent bi-phasic state of the pituitary, that the second phase of perphenazine action predominantly involves the biosynthesis of the hormone.

Activation of the pituitary by perphenazine appears to be similar to that obtained by pituitary transplantation (BOOT, ROEPCKE and MUEHLBOCK, 1964), and found in pituitary isografts in rats (NIKITOVITCH-WINER and EVERETT, 1958) and mice (BOOT et al., 1964). These isografts induced mammary differentiation (BARDIN, A. G. LIEBELT and R. A. LIEBELT, 1962; DAO and GAWLAK, 1963), and lactation in the host animals (COWIE, TINDAL and BENSON, 1960; HARAN-GERA, personal communication). Activation in both cases has been interpreted as being due to the elimination of the inhibitory effect of PIF by spatial separation of the hypothalamus from the pituitary. As pointed out above, perphenazine can similarly be considered a blocking agent of the PIF.

In summary, treatment with perphenazine results in a prompt release of prolactin from the pituitary into the blood, followed by a rate of production and release of the hormone which first increases and later levels off at a stage significantly above that of the controls. The elevated steady-state production and release of the hormone is accompanied by increased weight of the pituitary and by mammary gland growth and secretion.

g) Intracerebral Implantation of Phenothiazine Derivatives

In order to locate the mammotropic effect of phenothiazine derivatives, intracerebral stereotaxic implantations were carried out by MISHKINSKY, LAJTOS and SULMAN (1966).

Thirty-eight adult female rabbits weighing 3.5—4.5 kg were ovariectomized and primed with estradiol (0.1 mg s.c. daily) for 10 days. On the 11th day the left abdominal mammary gland was removed under pentobarbital anesthesia, and a stainless steel tube, bearing a perphenazine crystal, was lowered stereotaxically into different regions of the brain: the frontal lobe (6 animals), the thalamus (5 animals), the occipital lobe (5 animals), the median eminence (7 animals), and the pituitary anterior lobe (5 animals). In the control experiments, empty tubes were similarly implanted into the median eminence (5 rabbits); in addition, tubes bearing the phenothiazine derivative aminopromazine (Lispamol) which does not initiate lactation were inserted into the same site (5 rabbits). The implants were prepared by heating small perphenazine crystals (about 10 mg) on the tip of a 0.4 mm-diameter stainless steel tube. The tube was then thoroughly cleaned, leaving only the small perphenazine crystal visible on it.

Implantation into the median eminence was achieved by inserting the tubes at an angle of 6° to the midline in order not to damage the superior sagital sinus. All other implantations were carried out vertically on either side of the midline, at a distance of 1.5 mm. After implantation, the tube was fixed to the skull of the rabbit with dental cement and 2 anchoring screws.

On the 16th day the rabbits were killed and their abdominal mammary glands taken for autopsy, stained with hematoxylin and eosin, examined and graded on the mammotropic index scale (MTI) of 1—5, used in our Department (Chapter 26). At autopsy, the brains were carefully dissected and fixed in 4% formaldehyde solution and the steel tubes removed. They were then sectioned by the paraffin method into 7 μ slides and stained with hematoxylin and eosin. The localization of the implantation sites (Fig. 26) was verified from the serial sections.

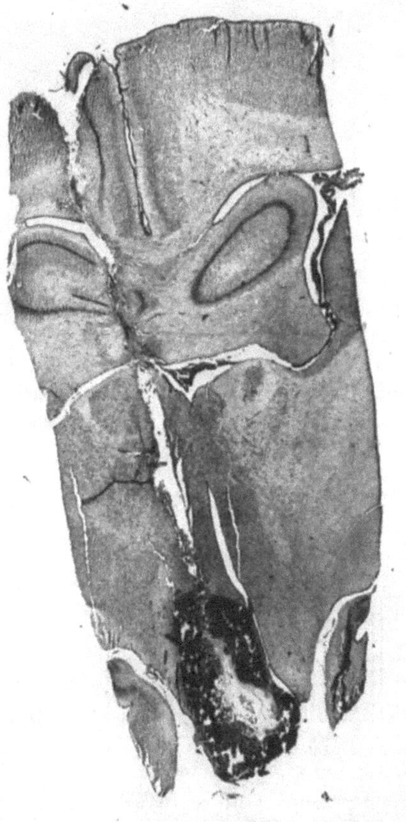

Fig. 26. Transverse frontal section of rabbit hypothalamus at P_1, showing an implanted site of the posterior basal median eminence at the bottom. The implant (black) promoted prolactin secretion, empty tubes (control) did not

The results are summed up in Fig. 27. No mammary development was noticed in rabbits with perphenazine implanted into their frontal lobe, thalamus, occipital lobe or pituitary. In a few cases we observed mammary involution, but full development could be observed only where perphenazine had been implanted into the median eminence of the hypothalamus (MTI 5). These implantations increased mammary gland development and brought it to the maximum stage Vb of the rabbit MTI scale. Neither aminopromazine (Lispamol) (Fig. 28) nor empty tubes evoked any mammary development.

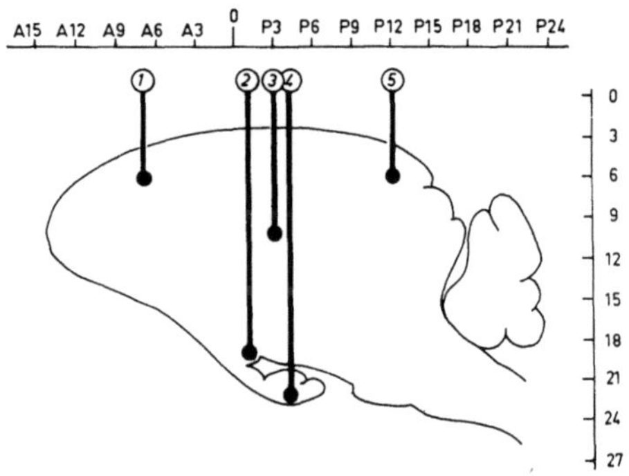

Fig. 27. Sites of perphenazine implantations in rabbit brain: *1* frontal lobe; *2* median eminence; *3* thalamus; *4* pituitary anterior lobe; *5* occipital lobe. (A, P, H are the stereotaxic coordinates.) Only median eminence implantation (*2*) promoted mammary development

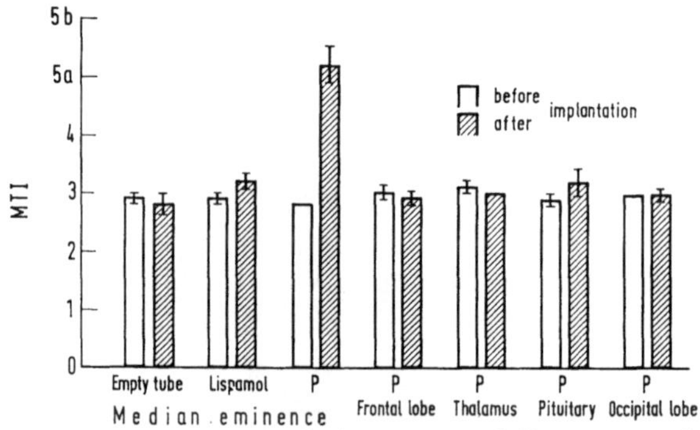

Fig. 28. Average mammotropic index (MTI) following implantation of perphenazine (P) crystals into different sites of the brain of ovariectomized, estrogen-primed rabbits (0.1 mg per day for 10 days). Only perphenazine implantation in the median eminence produces lactation. Lispamol (aminopromazine)—a non-lactogenic phenothiazine derivative—gives a negative reaction, as does the empty tube control (cf. also Chapters 14 and 19)

These results indicate that in rabbits the neurons of the posterior basal tuberal region of the hypothalamus are involved in the inhibitory mechanism which governs prolactin secretion by production or release of a PIF. Perphenazine implantations, like electrolytic lesions, apparently decrease the PIF level of the hypothalamus, causing release of prolactin from the anterior pituitary and activation of the mammary glands. It is assumed that perphenazine interferes with a balance established within the hypothalamus between the PIF and PRF mechanisms, by eliminating the inhibitory one. This interference follows the same rules of specificity as those established in section f of this chapter, since no mammary development was obtained when aminopromazine implants of the same diameter as that of the perphenazine-bearing electrodes were used for control. These results, thus, well confirm the hypothesis that a PIF is located in the posterior basal tuberal region of the hypothalamus.

h) Mammotropic Effect of Fluphenazine Enanthate in the Rat

Fluphenazine enanthate is, to date, the main long-acting phenothiazine tranquilizer widely used by psychiatrists. Laboratory experiments (Ebert and Hess, 1965) and clinical experience (Kinross-Wright and Charalampous, 1965) suggest that its tranquilizing effect lasts for 10—28 days with few side effects. Lately it has been shown that both fluphenazine HCl (Khazen, Mishkinsky, Ben-David and Sulman, 1968 a) and fluphenazine enanthate (Ben-David, 1968 a) evoke mammotropic activity, which appears to be one of their main side effects. Mishkinsky, Khazen and Sulman (1969), therefore, studied mammary development in rats for thirty days after subcutaneous injection or hypothalamic implantation of fluphenazine HCl and fluphenazine enanthate, in order to elucidate the mechanism of action of these two substances.

We used female albino rats weighing 200 ± 10 g each. They were primed for 10 days s.c. with 8 μg/rat/day estradiol. On the 11th day the animals were divided into five groups and injected s.c. as follows: In groups 1 (35 rats) and 2 (28 rats), each animal received a single injection of fluphenazine HCl, 5 or 25 mg/kg, respectively. In groups 3 (35 rats) and 4 (29 rats), each animal received a single injection of fluphenazine enanthate, 5 or 25 mg/kg, respectively; the animals in group 5 (35 rats) received a single injection of 0.2 ml sesame oil, and served as controls. A further 98 rats were primed for 10 days as above, and were implanted on the 11th day stereotaxically in the median eminence of their hypothalamus with glass tubings (Rudmond's disposable pipettes), bearing the following substances: group 6 (25 rats)—fluphenazine base; group 7 (27 rats)—fluphenazine enanthate; and group 8 (36 rats)—the control group—pure cocoa butter. The fluphenazine base and fluphenazine enanthate were melted prior to the experiment in cocoa butter at a ratio of 1:4 by weight, and

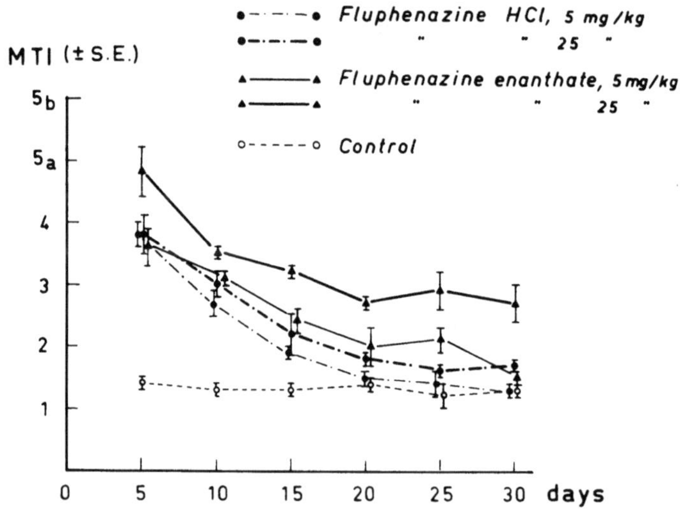

Fig. 29. Duration of mammotropic reaction after s. c. injection of fluphenazine salt compared with that produced by its ester. Average mammotropic index (MTI) of estradiol-primed rats, followed up for 30 days after single s. c. injections of fluphenazine HCl and fluphenazine enanthate

were tamped into the glass tubing. Subsequently, the external surface was carefully cleaned and the excess material removed. About one-sixth of the animals in each of the 8 groups were sacrificed at 5-day intervals from the 5th day until the 30th day after treatment and their MTI determined.

From Fig. 29 it can be seen that single injections of fluphenazine HCl and fluphenazine enanthate provoked moderate mammotropic activity, the MTI being about 4 on the 5th day after the single injection of either of the two substances and declining, after the 15th day, to MTI 1.7 in the fluphenazine HCl-injected rats and to MTI 2.7 in the fluphenazine enanthate-injected rats. The slope of mammotropic involution was not quite as steep in the rats which received fluphenazine enanthate as in those which received fluphenazine HCl. Implants of fluphenazine base caused a significantly higher mammotropic effect than those of fluphenazine enanthate (Fig. 30). Moreover, there was a tendency towards involution of the developed mammary pads 15 days after fluphenazine base implantation, at a time when fluphenazine enanthate developed its mammotropic effect lasting up to the 30th day. Beyond this date its effect declined.

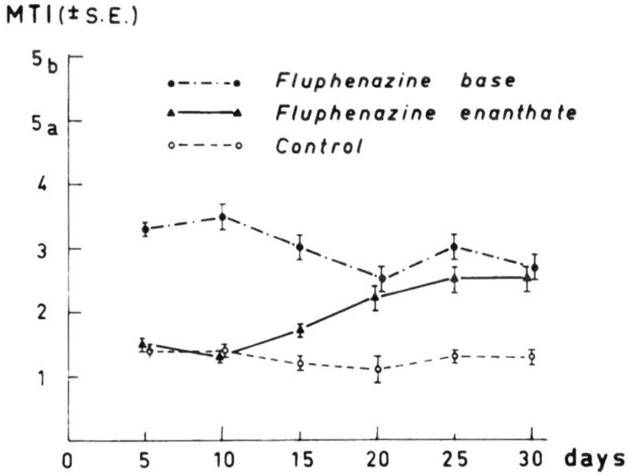

Fig. 30. Development of mammotropic effect after stereotaxic implantation in the hypothalamic median eminence. Average mammotropic index (MTI) of estradiol-primed rats, followed up for 30 days after stereotaxic implantation of 0.6 mg fluphenazine base and fluphenazine enanthate. Note the delayed and inferior mammotropic reaction elicited by the enanthate

Fluphenazine enanthate—the heptanoic acid ester of fluphenazine—was developed as a long-acting phenothiazine derivative to control severe cases of schizophrenia by monthly or bi-monthly injections. Although all phenothiazines have side effects, and fluphenazine enanthate, like other phenothiazine derivatives, may produce extrapyramidal reactions, Parkes, Brown and Monck (1962) showed that only a few of the patients receiving fluphenazine enanthate injections suffered any side effects—mainly drowsiness, lethargy, blurred vision or hypotension. Metabolic or endocrine side effects, such as weight-gain or lactation, have only rarely been reported.

Our interest in the mechanism of action of fluphenazine enanthate was aroused by previous studies which had established the structure-activity relationship of phenothiazines (KHAZEN, et al., 1968 a) and phenothiazine-like substances (KHAZEN, MISHKINSKY, BEN-DAVID and SULMAN, 1968 b). We concluded that there was a possible hypothalamic site sensitive to the phenothiazine moiety, and the question arose whether esterification would impair its lactogenic activity. It emerged during the 30-day follow-up that fluphenazine enanthate has a definite mammotropic activity when injected, but less so when implanted in the brain, probably due to its inability to undergo de-esterification in that organ. These results are supported by the findings of EBERT and HESS (1965), who demonstrated that fluphenazine enanthate metabolized in the rat into fluphenazine and fluphenazine sulphoxide, but that only free fluphenazine was to be found in the brain.

The implanted quantity of both fluphenazines (0.6 mg base per rat) was rather small, yet sufficient to yield a clear mammotropic effect without sedation. EBERT and HESS (1965), injecting higher doses (30 mg/kg), showed in their drug-metabolism study that only 0.2% of the radioactive fluphenazine enanthate was found in the brain immediately after injection, and this decreased to 0.04% after 21 days. This dose was sufficient to produce tranquilization and inhibit conditional avoidance responses for the whole three-week period, the animals reverting to normal thereafter. This led EBERT and HESS (1965) to conclude that the active agent ultimately deposited in the brain was fluphenazine base, a hypothesis which is confirmed in the present investigation by stereotaxic implantation of the two substances.

It seems that the slowing-down of the sedative effect parallels that of the mammotropic effect, supporting the view of BEN-DAVID (1968 a) that a persistent influence of the drug must be maintained for a minimum of 4 days in order to induce artificial hypothalamic lactation.

i) Comparative Study of Phenothiazine Effect on Hypothalamic Prolactin-Inhibiting Factor and MSH-Releasing Factor

It has recently been reported that acid extracts of rat hypothalamus, injected intravenously into recipient rats, induce an acute decrease in the melanocyte-stimulating hormone (MSH) concentration in the pituitary. The active agent in the hypothalamus has been referred to as a melanocyte-stimulating hormone-releasing factor (MSH-RF). It has been shown to be effective at low doses, the degree of the effect being related to the dose injected (TALEISNIK, ORIAS and OLMOS, 1966).

Nine phenothiazine derivatives—fluphenazine, trifluoperazine, perphenazine, thiopropazate, chlorpromazine, triflupromazine, methoxypromazine, prochlorperazine, promazine—were found to cause melanophore dispersion in the intact frog; mepazine was shown to be ineffective. When tested in hypophysectomized animals, none of the drugs showed any effect. The release of MSH is effected by direct action on the hypophysis or indirectly via the hypothalamus (SCOTT and NADING, 1961).

It is striking that the same phenothiazine derivatives which are most active in releasing prolactin are also potent MSH releasers. *In vitro* experiments on the mechanism of prolactin and MSH release by phenothiazine derivatives have recently been started in our Department.

Pituitaries and hypothalami were taken from normal ewes at the slaughter house within 5 min of slaughtering and put on ice. The organs were thawed immediately before the experiment, and frontal slice sections were prepared from the pituitary (1.5 mm thick) and from the hypothalamus (3 mm thick). One-half of a pituitary slice (about 100 mg) was cut into ten small fragments and distributed on 5 small Erlenmeyer flasks, containing 2 ml of medium M-199 each. The hypothalamic slices (about 300 mg each) were cut into 30 pieces and distributed among five other Erlenmeyer flasks containing two fragments each from the second half of the pituitary slice. After 4 hrs. shaking in a Dubnoff shaker at 37° C in a 5% CO_2/O_2 atmosphere, both sets of flasks were tested for prolactin and MSH effect. Prolactin concentration was determined with the local pigeon crop sac test at various dilutions, and MSH was tested on the tree frog Hyla arborea (SULMAN, 1952, 1956 a).

Our preliminary results show that the hypothalamus of the ewe inhibits prolactin release from its pituitary but enhances the production of MSH. These experiments will now be extended in order to find a method of separating and concentrating the two hypothalamic factors involved. Concurrently, mediators, releasers and inhibitors of this reaction will be studied *in vivo* and *in vitro*.

j) Conclusions

Screening of phenothiazine derivatives and their sulfoxides shows that both are lactogenic if the chemical structure includes a straight three-membered carbon side-chain at N_{10}. Sulfoxide formation nearly abolishes the sedative effect (down to 5%), without affecting lactogenic activity, operating via depression of the hypothalamic PIF. Prolactin is thus released from the pituitary.

Proof of this could be provided by direct assay of prolactin in blood and its depletion from the pituitary. Another approach is the stereotaxic implantation of phenothiazine derivatives into the hypothalamus. Derivatives, which are active when injected s.c., are also active by implantation; inactive derivatives are as negative as "empty" implants. The center which reacts to such implantation is located in the tuberal region of the posterior hypothalamus (median eminence). Fluphenazine enanthate is of special interest since its hypothalamic implantation is less lactogenic than implantation of fluphenazine base. This is due to the fact that it cannot enter the "receptor" of the PIF region unless broken down into unesterified fluphenazine. After subcutaneous injection, however, fluphenazine enanthate is de-esterified locally and has higher lactogenic activity than fluphenazine hydrochloride.

The structure-activity relationship (SAR) rules pertaining to the lactogenic effect of phenothiazine derivatives hold also for their melanophore activity in frogs. This points to the existence of a similar hypothalamic receptor for PIF and MIF.

Chapter 15

Mammotropic Effect of Phenothiazine-like Compounds

In Chapter 14 we have shown that a 3-C side chain—which may structurally form half a hexane ring—on a heterocyclic nucleus is far more important for the mammotropic effect than the ring system as such. This would explain why the phenothiazine nucleus by itself is not mammotropic (Khazen, Mishkinsky, Ben-David and Sulman, 1968 a). It has also been established that sulfoxidation of a phenothiazine nucleus minimizes its tranquilizing effect without substantially affecting its lactogenic potency (Ben-David, Dikstein and Sulman, 1965; Khazan, Ben-David, Mishinsky, Khazen and Sulman, 1966). Modern psychiatry tends towards use of phenothiazine-like substances in preference to the classic phenothiazines, because of the latter's chronic side-effects, i. e. photo-sensitivity, liver damage, eye melanosis, etc. Many related heterocyclic systems have, therefore, been synthetized, and these were tested by us.

Since the first synthesis of phenothiazine by BERNTHSEN in 1883, many modifications have been made in its nucleus, mainly by replacing the sulphur and nitrogen bridges between the two benzene rings by methylene or ethylene groups. STEINER and HIMWICH (1963), in their review on modified phenothiazine systems and imipramine derivatives, concluded that substitutions in one or both bridges between the benzene rings of phenothiazine tranquilizers reduce the antidepressant effect. STEINER's subjects were humans, and his parameters for tranquility were based on subjective observations, since the EEG pattern after administration of phenothiazines did not differ from that induced by phenothiazine-like substances. This general finding was confirmed both for phenthiazine sulfoxides (KHAZEN, MISHKINSKY, BEN-DAVID and SULMAN, 1968 a) and for the modified phenothiazines (KHAZEN, MISHKINSKY, BEN-DAVID and SULMAN, 1968 b).

The striking result, i. e. that the unsubstituted phenothiazine nucleus has no mammotropic effect, led us to investigate other similar nuclei, the main groups tested being the thioxanthenes and the iminodibenzyls.

KHAZEN et al. (1968 b) screened 18 ring-systems related to phenothiazine (Fig. 31), all of which contain a system of three consecutive rings: Rings A and C are mainly benzenic (with three exceptions), whereas ring B is composed of six or seven components. We found that, while the phenothiazine system is not required for achieving a mammotropic effect, and that a straight side chain—consisting of 3 carbons—may compensate for adverse changes in the phenothiazine nucleus, profound changes in it drastically reduce the mammotropic effect. Such substances may therefore find wider use only in psychiatry (LAUNCHBURY, 1964).

The mammotropic index and body temperature drop were evaluated as in Chapter 14. The results which are compiled in Table 3 show that the phenothiazine-like chemicals tested are less mammotropic but also less sedative than the corresponding phenothiazines. Two groups—thioxanthenes and iminodibenzyls—furnished some slightly mammotropic derivatives, but their low MTI proved that sulphur and nitrogen bridges in ring B are vital for a higher mammotropic effect. The most mammotropic phenothiazine-like substances were

Table 3. *Mammotropic effect of 52 phenothiazine-like compounds in oestrogen-primed female rats (200 ± 10 g). Oestrogen priming: 8 μg/rat/day for 10 days; treatment: 5 and 40 mg/kg/day for 5 days. Preparations are listed in descending order of their lactogenic effect at 5 mg/kg*

Generic and proprietary name	Structure: Nucleus type (Fig. 31)	Structure: R_1	Structure: R_2	Daily dose mg/kg	Mammotropic index (MTI)	Body temperature drop °C
P-4657 B (Pfizer)	1	$-SO_2N(CH_3)_2$	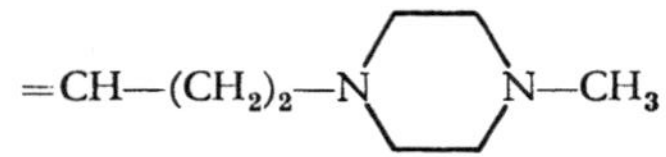	5 40	4.2 4.4	0.1 0.4
Clopenthisol (Ciatyl, Tropon) (Sordinol, Lundbeck)	1	—Cl	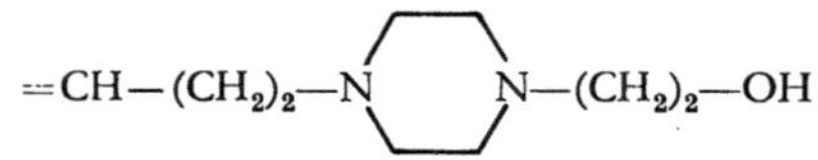	5 40	3.3 4.4	1.8 3.5
Chlorprothixene (Taractan, Roche)	1	—Cl	$=CH-(CH_2)_2-N(CH_3)_2$	5 40	3.0 3.0	1.7 4.8
Sandoz 11—18	1	—H	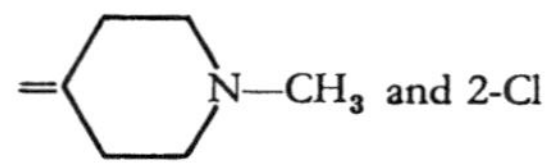	5 40	2.6 3.9	1.0 2.3
Sandoz BP 400	1	—H	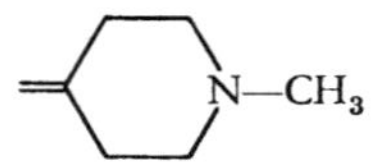	5 40	2.6 2.9	0.1 1.2
Methixene (Tremaril, Wander)	1	—H	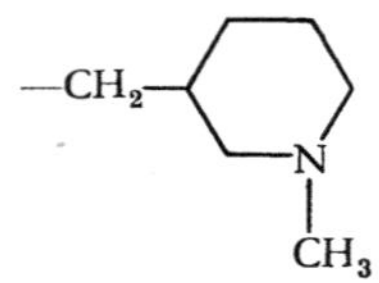	5 40	2.6 2.0	0.7 0

Sandoz 11—665 (BP 400 Sulfoxide)	1	—H	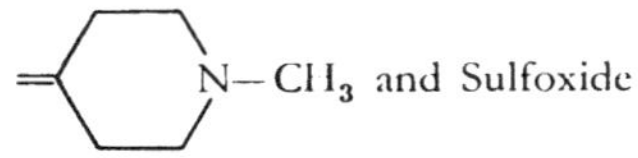	5 40	2.4 2.6	0.1 1.2
Sandoz 10—798	1	—H	$=CH—CH_2—N(CH_3)_2$	5 40	2.2 2.3	0 0.6
Sandoz 11—445	1	—H	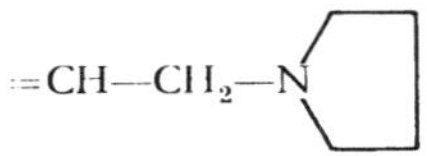	5 40	1.4 2.2	0.2 0.2
Oxyperpendyl (Pervetral, Homburg)	2	—H	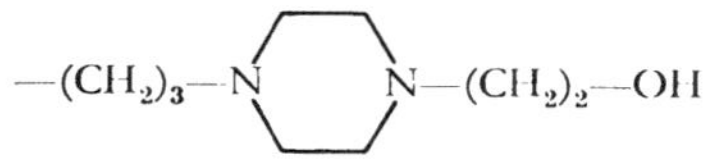	5 40	3.3 3.7	1.2 7.5
SQ-15283 (Squibb) (Selvigon, Homburg)	2	—H	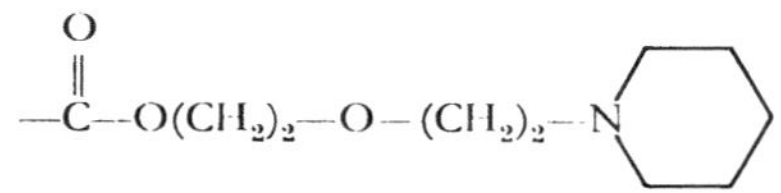	5 40	2.9 2.2	0 0
Isothipendyl (Andantol, Homburg)	2	—H	$—CH_2—CH(CH_3)—N(CH_3)_2$	5 40	2.8 2.2	1.3 1.7
Prothipendyl (Dominal, Homburg)	2	—H	$—(CH_2)_3—N(CH_3)_2$	5 40	1.8 1.7	2.1 9.5
Sandoz 12—039	3	—H	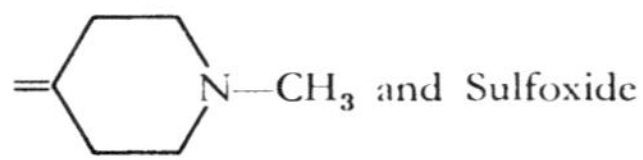	5 40	2.4 2.2	0.3 2.0

Table 3 (continued)

Generic and proprietary name	Structure: Nucleus type (Fig. 31)	Structure: R_1	Structure: R_2	Daily dose mg/kg	Mammotropic index (MTI)	Body temperature drop °C
SKF 9062-A_2 (Smith, Kline & French)	4	$—CF_3$	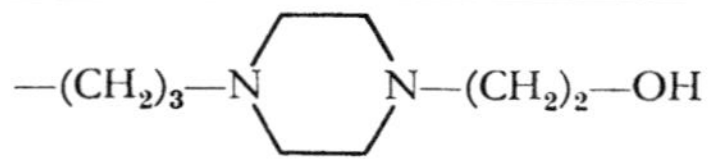	5 40	2.0 3.7	0 0.1
Brilliant Cresyl Blue (National Aniline)	4	$—CH_3$	—H and 8=NH, 2—$N(CH_3)_2$	5 40	1.6 2.4	0 1.2
Gallocyanin (I.G. Farben)	4	—H	—H and 1—OH, 2=O, 4—COOH, 8—$N(CH_3)_2$	5 40	1.6 2.5	0.6 0
Eosin Blue (Edward Gurr)	5	$—NO_2$	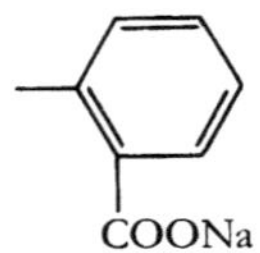 and 1,9—Br and 2,8=O and 7—NO_2	5 40	1.9 2.2	0.8 0.7
Methantheline (Banthine, Searle)	5	—H	—C(=O)—O—$(CH_2)_2$—$N(C_2H_5)_2$	5 40	1.7 2.4	0.9 1.9
Merbromine (Mercurochrome, Dobler)	5	—Br	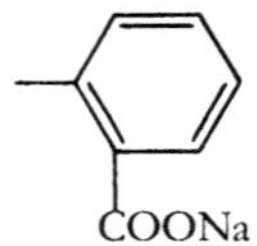 and 1—HgOH and 2—ONa and 8=O and 7—Br	5 40	1.2 3.3	0 0

Quinacrine (Atabrine, Winthrop)	6	—Cl	$-N-CH-(CH_2)_3-N(C_2H_5)_2$ H CH_3 and 7—OCH_3	5 40	2.0 3.2	0 0
Acridon (Fluka)	6	—H	—H and 9=0	5 40	2.0 1.8	0.3 0.9
Proflavine (BDH)	6	$-NH_2$	—H and 6—NH_2	5 40	1.7 2.4	0.1 1.5
Acriflavine (Pfizer)	6	$-NH_2$	$-CH_3$ and 6—NH_2	5 40	1.5 3.3	0.8 6.4
Litracen (N 7049, Lundbeck)	7	—H	$=CH(CH_2)_2-NHCH_3$ and 9,9-dimethyl	5 40	2.4 3.1	0.8 0.3
Melitracen (N 7001, Lundbeck)	7	—H	$=CH-(CH_2)_2-N(CH_3)_2$ and 9,9-dimethyl	5 40	2.2 2.4	0 0
Alizarin (National Aniline)	7	—OH	=O and 2—OH, 9=O	5 40	1.4 2.2	0.6 1.0
Trimepramine (Surmontil, Specia)	8	—H	$-CH_2-CH-CH_2-N(CH_3)_2$ CH_3	5 40	2.4 3.5	0 0.6
Imipramine (Tofranil, Geigy)	8	—H	$-(CH_2)_3-N(CH_3)_2$	5 40	2.4 2.8	0 2.7
G-34586 (Anafranil, Geigy)	8	—Cl	$-(CH_2)_3-N(CH_3)_2$	5 40	1.8 1.9	0 0.4

Table 3 (continued)

Generic and proprietary name	Structure: Nucleus type (Fig. 31)	Structure: R_1	Structure: R_2	Daily dose mg/kg	Mammotropic index (MTI)	Body temperature drop °C
R.P. 8307 (Specia)	8	—H	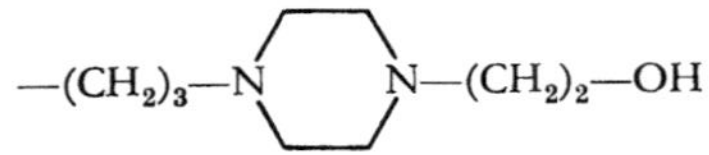	5 40	2.2 2.8	0 0.3
G-22150 (Geigy)	8	—H	$—CH_2—CH(CH_3)—N(CH_3)_2$	5 40	2.0 2.8	0.4 0.6
G-32883 (Geigy)	8	—H	$—CONH_2$	5 40	1.9 2.6	0 0.7
GP-33006 (Geigy)	8	—H	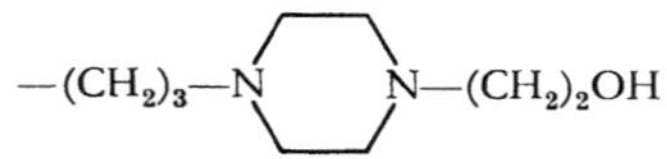	5 40	2.7 3.5	0.1 0.4
Nor-Imipramine (Pertofran, Geigy)	8	—H	$—(CH_2)_3—NHCH_3$	5 40	1.4 2.9	0.4 2.7
Carbamazepine (Tegretol, Geigy)	9	—H	$—CONH_2$	5 40	2.2 1.8	0 2.9
G-31406 (Geigy)	9	—H	$—(CH_2)_3—H(CH_3)_2$	5 40	2.2 3.2	0 1.1

Opipramol (Insidon, Geigy)	9	—H	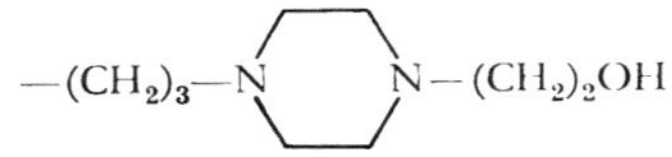	5 40	1.5 2.0	0 0.7
Deptropine (Brontine, Brocades)	10	—H		5 40	1.6 2.0	0.4 2.4
Nortriptyline (Aventyl, Lilly)	10	—H	$-(CH_2)_3-NHCH_3$	5 40	1.6 1.6	1.3 2.2
Amitriptyline (Elavil, Merck) (Elatrol, Assia-Zori)	10	—H	$=CH-(CH_2)_2-N(CH_3)_2$	5 40	1.6 2.4	0.2 2.1
Cyproheptadine (Periactin, Merck)	11	—H	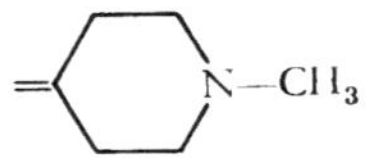	5 40	2.6 2.8	0 2.3
Protriptyline (Vivactil, Merck)	11	—H	$-(CH_2)_3-NHCH_3$	5 40	1.7 2.5	0.4 1.5
MK-130 (Merck)	11	—H	$=CH-(CH_2)_2-N(CH_3)_2$	5 40	2.4 4.4	0.1 1.7
Curatin (Doxepin, Pfizer)	12	—H	$=CH-(CH_2)_2-N(CH_3)_2$	5 40	2.0 1.2	0 1.1
AW-14,2333 (Wander)	13	—H	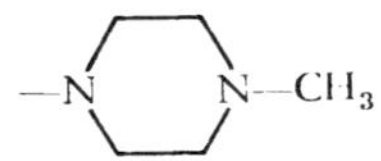	5 40	3.0 3.3	1.1 5.6

Table 3 (continued)

Generic and proprietary name	Structure			Daily dose mg/kg	Mammotropic index (MTI)	Body temperature drop °C
	Nucleus type (Fig. 31)	R_1	R_2			
SUM-3170 (Wander)	14	—Cl	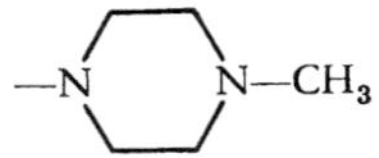	5 40	5.0 5.5	6.3 6.0
Dibenzodiazepine (Noveril, Wander)	15	—H	=O and 11—CH_3 and 5—$(CH_2)_2$—$N(CH_3)_2$	5 40	2.0 2.4	0.3 1.0
Clothiapine (HF-2159, Dorsey)	16	—Cl	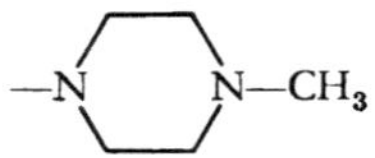	5 40	4.0 5.2	3.4 6.4
Clothiapine sulfoxide (self synthesis)	16	—Cl	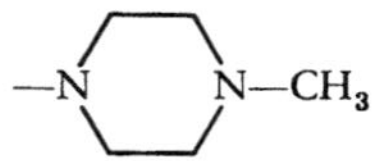	15	3.7	—
Sandoz BC 105	17	—H	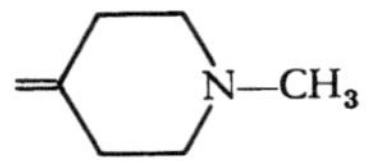	5 40	3.1 2.8	0.1 1.4
Sandoz 11—233	18	—H	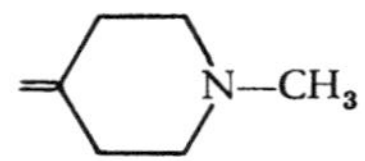	5 40	2.0 2.5	0.8 1.9
Saline (Control)	—	—	—	0.5 ml	1.3	0

P-4657 B and clopenthisol (thioxanthene nucleus), oxyperpendyl (azaphenothiazine nucleus), Sum-3170 (dibenzooxazepine nucleus) and clothiapine (dibenzothiazepine group).

It seems that the spheric orientation of the molecule is a most important factor for its effect on the central nervous system: the aza-

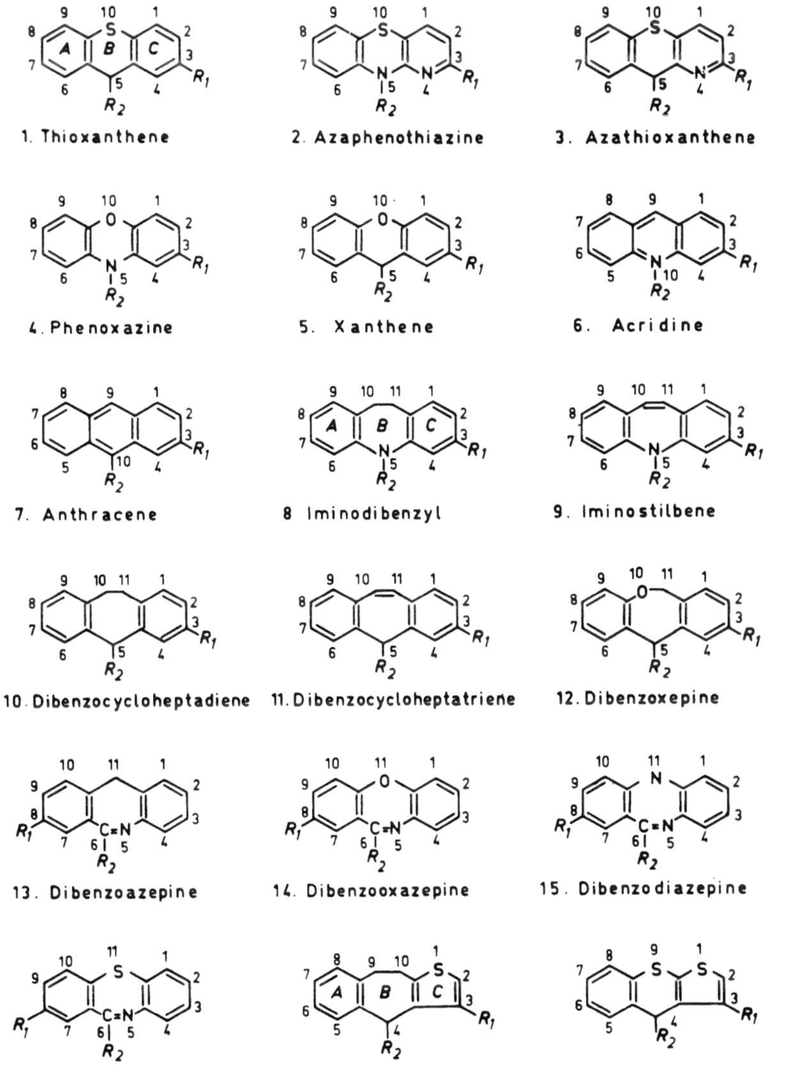

Fig. 31. Eighteen phenothiazine-like nuclei tested for their mammotropic effect. Only Nos. 1, 2, 14 and 16 yielded active compounds

phenothiazines containing nitrogen in the C ring had a markedly lower mammotropic effect. These findings suggest that chemicals bearing A and C benzene rings are the most active. It also seems that the mammotropic effect mediated by psychopharmaca requires a nitrogen atom in the nucleus: none of the substances lacking such a nitrogen atom had any mammotropic effect whatever.

Another important fact emerges from Table 3, *viz.* that there is no direct relationship between the mammotropic and the tranquilizing effects of the tricyclic nuclei tested, although in many cases both effects run somewhat parallel. It may be concluded that the diphenyl moiety is the minimum prerequisite for a mammotropic effect, and that phenothiazine is superior in this respect to all its variations. None of the derivatives tested had significantly high MTI values. The very popular derivatives of azaphenothiazine were non-mammotropic: oxyperpendyl, a substance closely related to perphenazine, was found to have a very slight mammotropic effect, due to lack of a polar group at C_3 and insertion of a second nitrogen in place of C_4. Much of the loss of the mammotropic effect in this case can be attributed to the spheric orientation of the nucleus bearing a second nitrogen.

The sulphur and nitrogen bridges may be replaced by carbons as long as the polar group at C_3 and the two benzene rings remain intact: dibenzothiazepine and dibenzooxazepine—when bearing a polar chlorine atom at C_3 and intact A and C benzene rings—are mammotropic, while substitution of the chlorine by a hydrogen atom markedly reduces the mammotropic effect. Neither the anthracene system nor the dibenzocycloheptanes were lactogenic: when the hetero-atom was shifted to the C ring (Table 3: BC-105), the substance had some mammotropic effect, but a very similar drug—cyproheptadine (Periactin)—without any hetero-atom had no mammotropic effect at all. This may be explained by the quick accumulation of the sulphur-containing rings in the brain (Steiner and Himwich, 1964). Thus it emerges that small changes in the three-carbon side chain, like demethylation of the terminal N, which proved to be very important for abolishing the mammotropic effect in phenothiazines, are of lesser importance in phenothiazine-modified substances.

In regard to sedation, Khazen et al. (1968 b) could in some instances confirm by body temperature drop the observation of Peterson and Nielson (1964), that central depression largely depends on the presence of a double bond between the nucleus and the R_2 side chain, and on polar substitution at C_3.

Another theory (ABOOD, 1955) relates the sedative effect of the phenothiazine-like substances to their reactivity with the phosphorylative complex of mitochondria in the hypothalamus. This complex greatly resembles the diphenyl moiety, due to its three-dimensional molecular structure. This may suggest that the phenothiazine-like substances, like many other diphenyls, achieve their depressive effect by inhibiting oxidative phosphorylation. This, however, does not hold true for the mammotropic effect.

To sum up this series of conclusions, one must admit that tricyclic variants of phenothiazine are no more lactogenic than the mother compound. Still, these compounds, although they help us considerably in understanding the mechanism of lactogenic drug action, do not supply us with a complete explanation. Only a precise receptor theory, which will be discussed in Chapter 19, can provide full understanding.

Chapter 16

Mammotropic Effect of Reserpine Derivatives

The endocrine effects of reserpine are well known. They are mediated by the hypothalamus (KHAZAN and SULMAN, 1961; KHAZAN, SULMAN and WINNIK, 1961; KHAZAN, BEN-DAVID and SULMAN, 1963).

Gynecomastia was reported by WILKINS (1954) to have developed in male patients following administration of Rauwolfia alkaloids. Rauwolfia alkaloids and reserpine were also reported by GAUNT et al. (1954, 1963) to interfere with estrus cycles in mice, and with ovulation and menstruation in monkeys (DE FEO and REYNOLDS, 1956). PLATT and SEARS (1956), WOOLMAN (1956), DURLACH (1957), MEITES (1957 a), ROBINSON (1957), TINDAL (1959), KHAZAN et al. (1962), KANEMATSU, HILLIARD and SAWYER (1963 a, b) and GATCHEW (1966) found reserpine to have a mammotropic and lactogenic effect in rats, rabbits and women. All these authors are agreed that prolactin hypersecretion and discharge from the anterior hypophysis are responsible for the mammotropic and lactogenic effects observed after administration of reserpine and phenothiazine derivatives. This has been confirmed by KHAN and BERNSTORF (1964) and was later also proved

Table 4. *Mammotropic effect of 16 reserpine analogues compared with their sedative effect*

Oestradiol priming: 8 μg/rat/day for 10 days s.c.
Reserpine treatment 0.5—20 mg/kg/day for 5 days s.c.
Female rats 200 (± 10) g
Scoring from 0 to 5:
Overt behaviour: 0 = alert → 5 = reserpine syndrome
Body temperature drop: 0 = 1/2 °C → 5 = 2.5 °C
Decrease in barbital sleep induction time: 0 = 0 min → 5 = 5 min
Mammotropic index: 1 = normal mammary gland
5 = lactating mammary gland

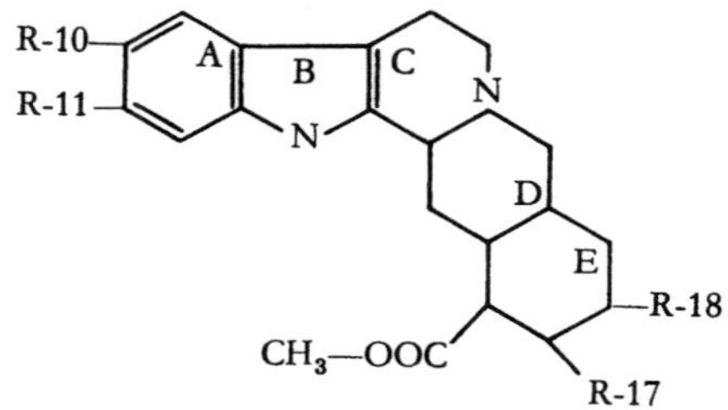

No.	Generic and proprietary name	R_{10}	R_{11}	R_{17}	R_{18}	Dose mg/kg	Scoring of sedative effect: Overt behaviour	Scoring of sedative effect: Body temperature drop	Scoring of sedative effect: Decrease in barbital Sleep induction time	Mammotropic index MTI	Recommendable for lactogenic use
0	Control: Saline solution	—	—	—	—	0.5—1 ml	0	0	0	1.5	—
1	Deserpidine Harmonyl-Abbott	—	—	OCH_3	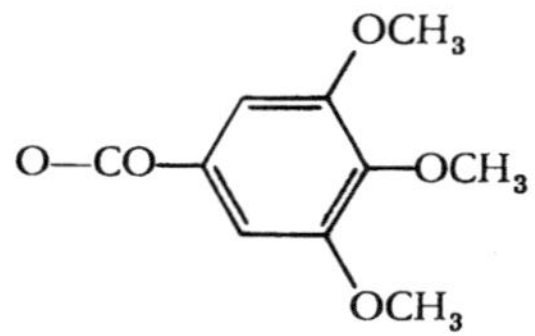	0.5 1	2.5 3.5	2.1 3.2	4.0 5.0	4.5 5.0	±
2	Reserpine Serpasil-Ciba	—	OCH_3	OCH_3	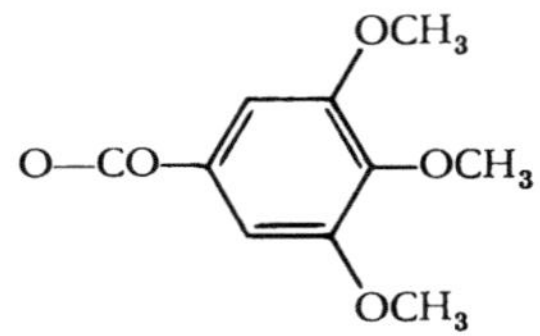	0.5 1	3.5 5.0	3.2 3.2	4.0 4.2	4.2 5.0	±

3	Rescinnamine Moderil-Pfizer	—	OCH_3	OCH_3	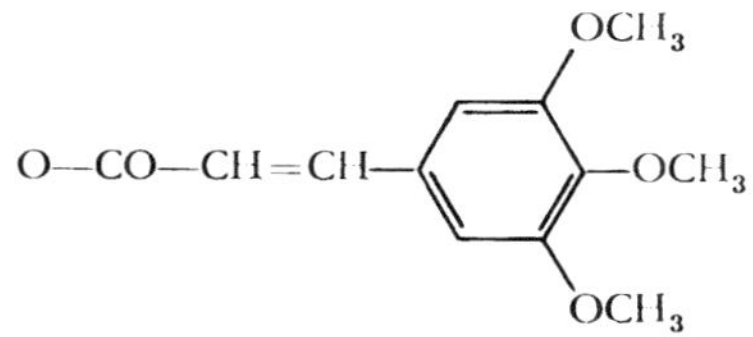	1 5	3.5 4.0	1.2 2.2	2.2 2.5	4.0 5.0	±
4	Su-10,704 Ciba (maleate)	—	OCH_3	OCH_3	$O—CH_2—CH_3$ in beta position	0.5 1	0 5.0	0 2.2	0 4.9	2.0 4.6	+
5	Su-9064 Ciba	—	OCH_3	OCH_3	$O—CH_3$ in alpha position	0.5 1 5 10	0 1.1 2.3 3.6	0 2.0 2.2 5.0	0 4.0 4.5 5.0	3.3 3.8 4.0 4.6	+
6	Tetrabenazine Nitoman-Roche	OCH_3	OCH_3	—	Ring B and E missing O and $CH_2—CH—(CH_3)_2$ on ring D	0.5 1 5 10	0 1 2.3 3.5	0 1 1.6 2.2	0 2 4.0 4.8	2.8 3.2 3.6 4.2	+
7	Su-7064 Ciba	—	OCH_3	OCH_3	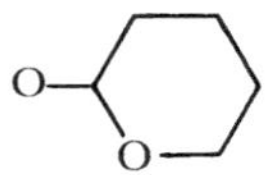 in beta position	0.5 10	0 3.1	0 2.2	0 2.3	2.8 3.6	+
8	Su-11,279 Ciba	—	OCH_3	OCH_3	$O—CH_2—CH_2—O—CH_3$. in alpha position at C_{16}: $CO—O—CH_2—CH_2—O—CH_3$	1 10 20	0 0 0	0 0 1.8	0 1.0 1.6	3.3 3.4 4.0	++
9	Su-9673 Ciba (Sulfate)	—	OCH_3	OCH_3	$O—CH_2—CH_2—CH_3$	0.5 1 5	0 0 2.5	0 0 2.1	0 1.3 2.6	1.8 2.3 3.4	±

Table 4 (continued)

No.	Generic and proprietary name	R_{10}	R_{11}	R_{17}	R_{18}	Dose mg/kg	Scoring of sedative effect: Overt behaviour	Scoring of sedative effect: Body temperature drop	Scoring of sedative effect: Decrease in barbital Sleep induction time	Mammotropic index MTI	Recommendable for lactogenic use
10	Su-10,092 Ciba (maleate)	—	OCH_3	OCH_3	O—CH_2—CH_2—CH_3	0.5 1 5	0 1 2.5	0 0 2.0	0 1.5 2.5	2.0 2.2 3.2	±
11	Benzquinamide Quantril-Pfizer	—	—	—	Ring B and E missing O—CO—CH_3 and CO—N$(C_2H_5)_2$ on ring D	5 10 20	1.5 2.3 2.5	0 1.1 1.6	1.6 3.2 4.0	2.6 3 3.2	—
12	Yohimbine Quebrachine	—	—	OH	—	1 5	1.1 1.5	1.4 4.2	2.2 3.7	2.2 2.6	—
13	Ajmaline Cardiorythmine-Servier	—	—	—	Ring E connected to ring C OH and C_2H_5 on ring D, CH_3 on ring B	1 5	0 2.3	0 1.1	1.2 2.5	1.6 2.6	—
14	Su-9300 Ciba (hydrochloride)	—	OCH_3	OCH_3	O—CH_2—CH_3 in alpha position	0.5 1	0 1.5	0 0	0 3	1.8 1.7	—
15	Methoserpidine Decaserpyl-Roussel	OCH_3	—	OCH_3	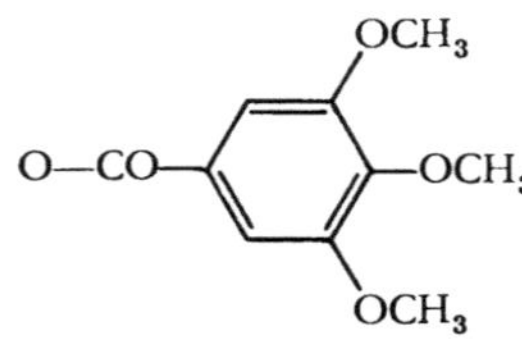	1 5	1.1 1.5	0 2.0	2.5 3.6	1.2 1.6	—
16	Ajmalicine Raubasin-C. H. Boehringer	—	—	—	C_{17} replaced by O C_{19} methylated C_{16}—C_{17}: Δ	1 5	1.1 1.5	0 2.0	0 3.6	1.8 2.2	—

by radioimmunoassay (BRYANT and GREENWOOD, 1968) and by bioassay (BEN-DAVID and CHRAMBACH, 1969).

KHAZAN, BEN-DAVID, MISHKINSKY, KHAZEN and SULMAN (1966) showed that dissociation between the mammotropic and sedative effects of tranquilizers is possible, both being caused by independent mechanisms. BEN-DAVID, KHAZEN, KHAZAN and SULMAN (1968) studied 16 reserpine derivatives (Table 4) for correlation of the endocrine and the psychopharmacologic properties. The analogues differed from each other through varying positions of the methoxy groups and different radicals attached to the etherial linkage at position 18 at both configurations alpha (epi) and beta (normal). One of the analogues studied—Su-11,279—was the only one belonging to the 3-iso-reserpate series. Solutions for injection were made with a non-toxic reserpine solvent (propylene glycol 25%, acetic acid 1.14%, aqua bidest. ad 100%) which could be injected with 0.5—1 ml of normal saline.

Four parameters were compared:

a) *The mammotropic effect* (MTI) was studied, for each dose level, in groups of 5 to 10 adult female rats, each weighing 200 ± 10 g. The animals were primed for 10 days by daily s.c. injections of 8 μg estradiol in oil. From the 11th to the 15th day, daily s.c. injections of the reserpine analogues were given, and on the 16th day the animals were sacrificed. The right inguinal mammary pads were removed and their MTI determined (scoring 1—5).

b) *The body temperature drop* (BTD) was studied in groups of 10 female rats (200 ± 10 g), which were injected s.c. with a single dose of the reserpine analogues. Rectal temperatures were recorded at 0, 1, 2, 3, 4 and 5 hours with an electrocouple thermometer (Electric Universal Thermometer TE_3 "Ellab Instruments", Copenhagen). Control animals were injected with 0.2 ml saline s.c. Scoring: 1 = −0.5 °C, 5 = −2.5 °C.

c) *The effect on general overt behavior* (GOB) was studied in groups of 10 female rats, weighing 125 ± 10 g each. Immediately after injection, the rats were put into individual jars to permit accurate evaluation of the effect. Ptosis was judged by the degree of dropping of the eyelid, the ability of the rats to reopen their eyes after shaking of the jars, and the time which elapsed till the rats closed their eyes again. Exploratory behavior was evaluated by putting the rats on an upside-down jar and measuring the time lapse till the rats calmed down. Irritability was judged by the rats' reaction towards pricking of their backs with a blunt hypodermic needle just above the attachment of the tail (Scoring 1—5, Fig. 32).

d) *The barbital sleep induction time* (BIT) was studied in groups of 10 female mice, weighing 30 ± 2 g each. Two hours after i.p. injection of the different analogues, the animals were injected with 500 mg/kg of barbital sodium. Control mice received barbital sodium alone. The time elapsing between barbital injection and hypnosis, measured in minutes, was considered the barbital induction time (BIT). An animal was judged to be hypnotic if it lost the righting reflex for one minute or longer. Scoring: 1 = 1.5 min., 5 = 8 min.

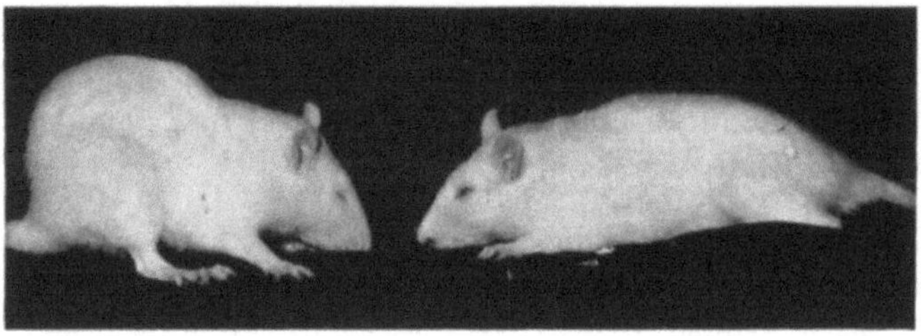

Fig. 32. Effect of reserpine and its derivatives on overt behavior of rats: l: reserpine syndrome (eye ptosis, arched position combined with irritability); r: normal control rat

Table 4 shows the comparative effectiveness of the reserpine analogues at different dose ranges used, arranged in order of their lactogenic activity (MTI: 1—5). At higher doses, all the reserpine analogues studied exerted strong sedative hypothalamic effect for less than 8 hours, except for Su-8,064 (10 mg/kg) and Su-10,704 (1 mg/kg) whose sedative effect lasted for more than 8 hours but was still shorter than the classical reserpine-syndrome which continues for 24 hours. Deserpidine, reserpine, rescinnamine and Su-10,704 were found to be the most effective analogues with regard to both their sedative and their mammotropic effects, when given at a 1 mg/kg dose level. Su-9,064, tetrabenazine, Su-7,064, Su-9,673, and Su-10,092 had weaker effects. Surprisingly, Su-11,279 was found to lack all sedative effect even at a dose as high as 20 mg/kg; on the other hand, it showed a high degree of mammotropic activity. Benzquinamide, yohimbine, ajmaline, Su-9,300 and methoserpidine had a very low lactogenic effect.

These results furnish an interesting example for the SAR between closely related chemical compouonds, comparable to earlier results with phenothiazine derivatives (cf. Chapter 14). It can be concluded that, when an ethyl radical in beta configuration is attached to position 18 on a reserpine molecule (Su-10,704), a most potent analogue will result, with regard to both sedative and mammotropic effects. If this ethyl radical is attached in alpha configuration (Su-9,300), the potency of the analogue drops sharply.

Attachment of a 2-methoxy-ethyl at position 18, in alpha configuration, to a 3-isoreserpate (Su-11,279) abolished the sedative effect, with the mammotropic effect remaining fairly high. This most interesting finding induced us to give Su-11,279 the highest scoring for practical use as a lactogenic agent. In general, scoring for recommendability of a substance for lactogenic use must be based on maximum lactogenic effect combined with minimal sedative effect. For this reason, Su-10,709, Su-9,064, tetrabenazine and Su-7,064 range higher than deserpidine, reserpine or rescinnamine in spite of the excellent lactogenic activity of the three latter substances.

Reserpine analogues with a methyl group in alpha or beta configuration were studied by SCHLITTER and PLUMMER (1964) for their sedative effects. The authors found that the beta-analogues had a higher sedative potency than the alpha-analogues. Highest potency was attained when an ethyl radical replaced the methyl in both positions. GALA and REECE (1963), in their study of Su-9,064, reported that this analogue was able to release lactogenic hormone from the anterior pituitary, similar to reserpine, without, however, being tranquilizing or hypotensive at the reserpine dosage. High doses were reported to induce the full reserpine-syndrome. The authors rightly concluded that the lactogenic effect of reserpine is not due to its hypotensive or tranquilizing action, but is inherent in the basic structure of the molecule.

PURSHOTTAM (1962) reported that syrosingopine, another reserpine derivative, had a hypotensive effect equaling that of reserpine, but was less sedative and less active in suppressing pituitary gonadotropin release. PALLADINO, BUTCHER and FUGO (1968) demonstrated that yohimbine suppressed the PIF and produced luteotropic hormone in pregnant rats. Whe showed that the mammotropic and the tranquilizing properties are unrelated, independent of each other, and mediated through two selective actions of the chemical agents on different CNS receptors, mediators or mechanisms (ASSAEL, GABAI, WINNIK, KHAZAN and SULMAN, 1960; KHAZAN, BEN-DAVID, MISHKINSKY, KHAZEN and SULMAN, 1966). Su-11,279 furnishes a classic example for such "dissociation," as it shows no sedative property but high mammotropic activity.

Summarizing it seems that reserpine and most of its derivatives have a strong lactogenic effect. The absence of all lactogenic effect in methoserpidine is particularly interesting because it differs from the highly lactogenic reserpine only in that the methoxy group is shifted from C_{11} to C_{10}. This fact shows the importance of a polar substitution at C_{11}. It would seem that the hypothalamus has a specific array of receptors which are sensitive to the action of cyclic compounds when they have a polar group at a corresponding spot, which is for the three-cyclic phenothiazines at C_2, for the four-cyclic steroids at C_3, and for the five-cyclic reserpine derivatives at C_{11}. Any compound fitting into this set-up is lactogenic but not necessarily tranquilizing, as has been shown. This problem will be further discussed in Chapter 19 devoted to receptor theories. Studies on the SAR of reserpine analogues supply further evidence for a possibility of complete dissociation between the mammotropic and sedative activities of hypothalamic tranquilizers.

Chapter 17

Mammotropic Effect of Butyrophenones

The clinical use of butyrophenones as neuroleptics originated at the research laboratories of Janssen Pharmaceutica in Beerse (Belgium) about ten years ago. It arose from a systematic investigation of the relationship between the chemical structure of 4-phenyl-piperidines and propiophenones and their pharmacological properties. Members of the higher homologue, the butyrophenones, proved to have a clear antipsychotic neuroleptic potency, and quite a few of them are used nowadays in psychiatry, obstetrics and anesthesiology (JANSSEN et al., 1959; HAASE and JANSSEN, 1965). They posses chlorpromazine-like properties (JANSSEN, 1967 a, b), and were brought to our attention through their mammotropic activity, which may be a possible side-effect at unusually high clinical dose levels.

Mishkinsky, Khazen, Givant and Sulman (1969) have so far tested 23 representatives of this group for their mammotropic activity —as measured by mammary gland development, their sedative effect— as measured by body-temperature drop, and their cataleptic effect— as measured by catatonic behavior and ptosis.

All parameters were scored by numbering them from 1 to 5, to allow easy comparison of effects, and the aggregate of the scores of each group was divided by the number of animals. The assay of the mammotropic index (MTI) is described in Chapter 26, and the assay of body-temperature drop (BTD) in Chapter 14. The other two parameters were assayed as follows:

Cataleptic Effect (CPE)

Cataleptic effect was determined in female rats weighing 150 ± 10 g (5 rats per drug per dose) 60 minutes after s.c. injection. The test procedure consisted of putting the rat gently on its hind legs, in a head-upward position, with one of its forelegs placed on cork plates 3 or 9 cm thick. In this test we followed the method of Wirth, Gösswald, Hörlein, Risse and Kreiskott (1958) who described different stages of cataleptic behavior in rats induced with phenothiazine compounds. The following five stages for the cataleptic effect were scored:

1. Transient cataleptic reaction lasting less than 30 seconds;
2. Cataleptic position, forelegs 3 cm raised for 30 seconds;
3. Cataleptic position, forelegs 9 cm raised for 30 seconds;
4. Prostrate posture with reaction to pricking;
5. Prostrate posture without reaction to pricking.

Failure to correct the imposed posture for 30 seconds was considered a positive cataleptic reaction. The test was carried out at each dose level on both right and left forelegs.

Ptosis (PTS)

Ptosis was evaluated by the degree of dropping of the upper eyelid. For scoring, the following five stages were used:

1. Intermittent lid closure;
2. One side semi-closed;
3. Both sides semi-closed;
4. One side completely closed;
5. Both sides completely closed.

Table 5. *Mammotropic index (MTI), body temperature drop (BTD), cataleptic effect (CPE) and ptosis (PTS) of 23 butyrophenones in oestradiol-primed female rats (200 ± 10 g). Oestradiol priming: 8 µg/rat/day for 10 days. Butyrophenone treatment: 1, 5 and 40 mg/kg/day for 5 days. Preparations are listed in descending order of their lactogenic effect (MTI) at 1 mg/kg*

Formula 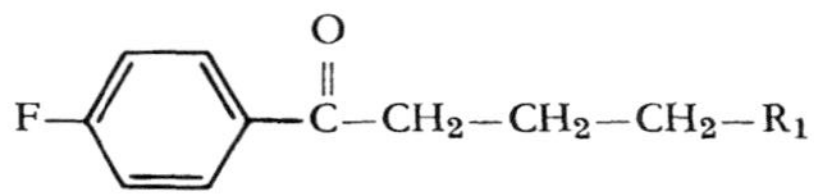

Code No. and generic anme	Substitution of R_1	Daily dose (mg/kg)	Mammotropic index (MTI)	Body temperature drop (BTD)	Cataleptic effect (CPE)	Eyelid ptosis (PTS)
R-2498 Trifluperidol (Janssen)	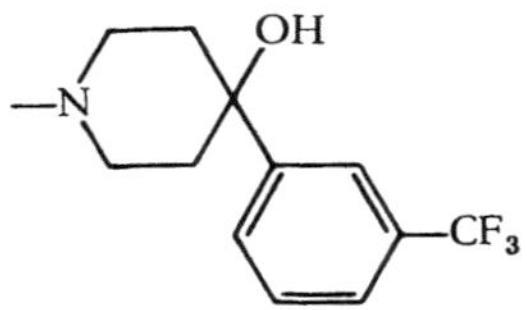	1 5 40	4.0 5.0 4.6	0.6 2.0 3.6	2.0 3.0 4.0	1.5 2.5 3.5
R-1658 Moperone (Janssen)	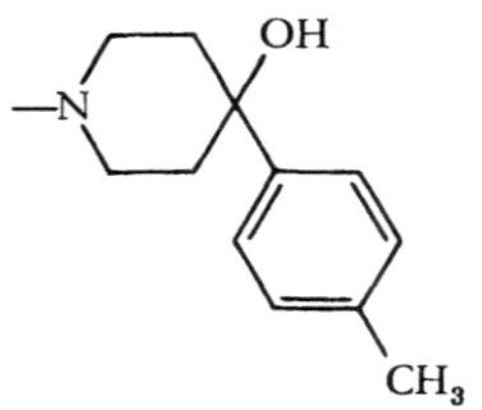	1 5 40	3.9 4.0 4.3	0.9 2.1 4.9	2.5 3.0 3.5	1.0 2.0 3.0

R-5147 Spiroperidol (Janssen)	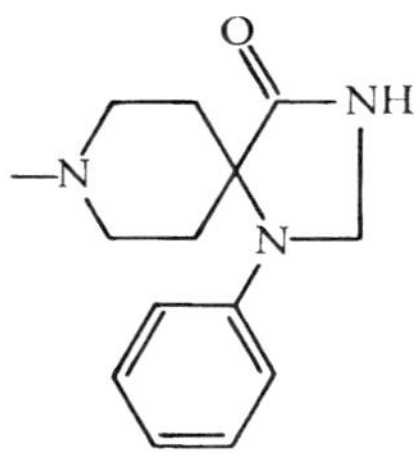	1 5 40	3.8 5.0 4.4	1.1 2.4 2.2	2.8 4.0 4.0	1.0 2.0 3.0
R-4749 Droperidol (Janssen)	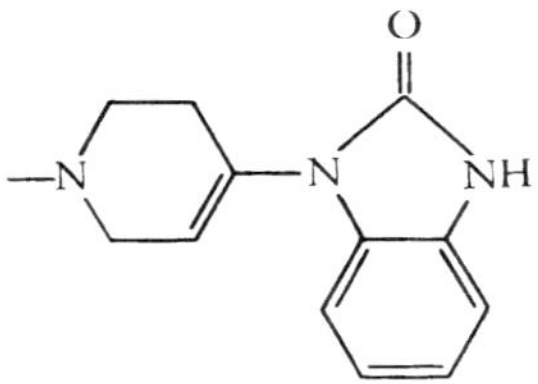	1 5 40	3.8 3.5 3.4	0.9 1.5 2.2	1.5 2.5 4.0	3.0 3.0 3.5
R-9298 Clofluperol (Janssen)	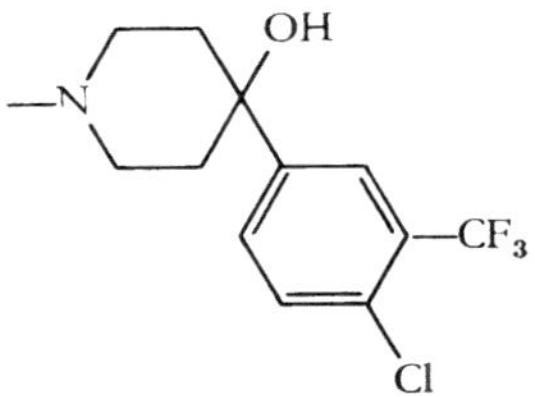	1 5 40	3.7 4.4 4.0	1.1 1.7 1.7	2.5 3.0 3.5	0 0.5 1.0

Table 5 (continued)

Code No. and generic name	Substitution of R_1	Daily dose (mg/kg)	Mammo-tropic index (MTI)	Body tem-perature drop (BTD)	Cataleptic effect (CPE)	Eyelid ptosis (PTS)
R-4584 Benzperidol (Janssen)	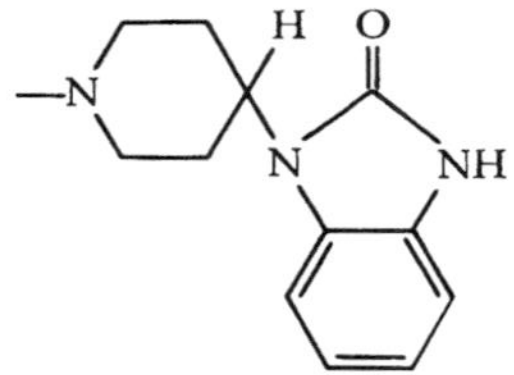	1 5 40	3.4 3.8 3.8	1.3 1.3 1.7	2.5 3.0 4.0	1.5 2.5 3.5
R-1625 Haloperidol (Janssen)	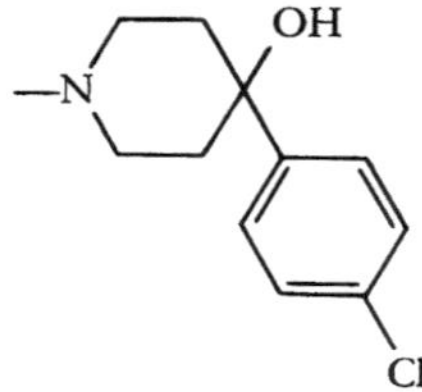	1 5 40	3.3 4.7 4.8	1.1 2.3 3.0	2.5 2.5 4.0	1.0 1.0 3.0
R-6238 Pimozide (Janssen)	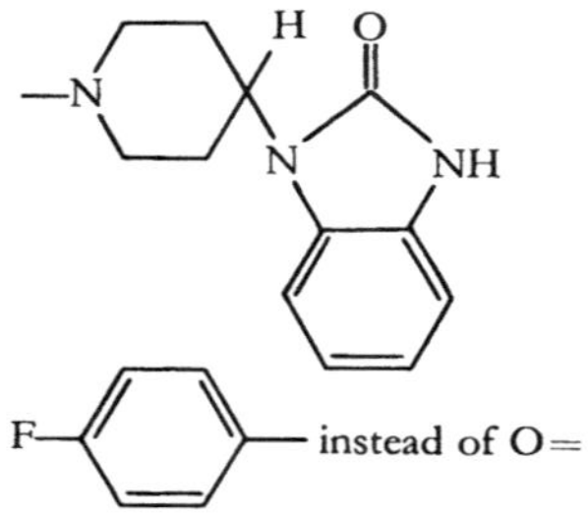	1 5 40	3.2 3.2 4.1	0.7 0.9 1.4	0 1.0 1.5	0 0 0.5

R-2028 Fluanisone (Janssen)	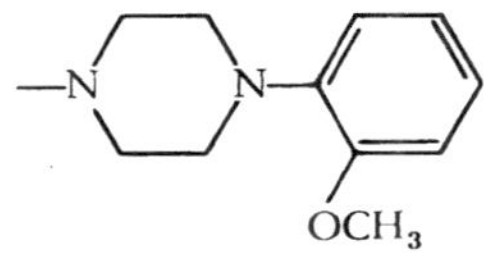	1 5 40	3.0 3.5 4.0	1.7 2.6 4.6	1.0 2.0 5.0	1.0 2.0 3.0
R-3201 Haloperidide (Janssen)	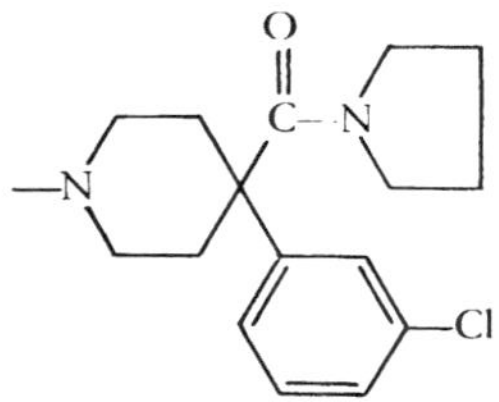	1 5 40	3.0 3.0 3.3	1.0 2.1 3.3	1.5 3.0 3.0	0.5 2.5 2.5
R-4457 (Janssen)	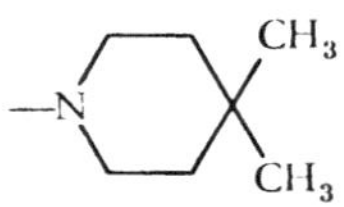	1 5 40	3.0 3.0 3.6	0.8 2.4 3.8	1.0 2.0 3.0	0 1.5 2.0
R-3345 Floropipamide (Janssen)	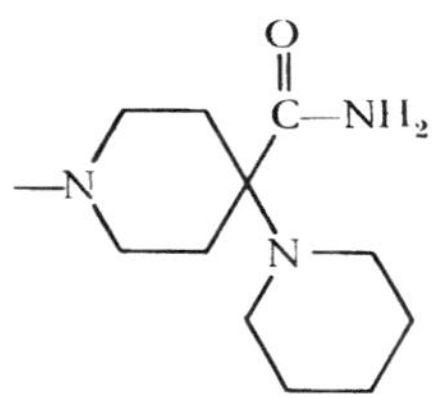	1 5 40	2.8 2.9 2.8	1.7 3.2 4.7	0 1.0 2.0	0 0 1.0

Table 5 (continued)

Code No. and generic name	Substitution of R_1	Daily dose (mg/kg)	Mammotropic index (MTI)	Body temperature drop (BTD)	Cataleptic effect (CPE)	Eyelid ptosis (PTS)
R-2963 Methylperidide (Janssen)	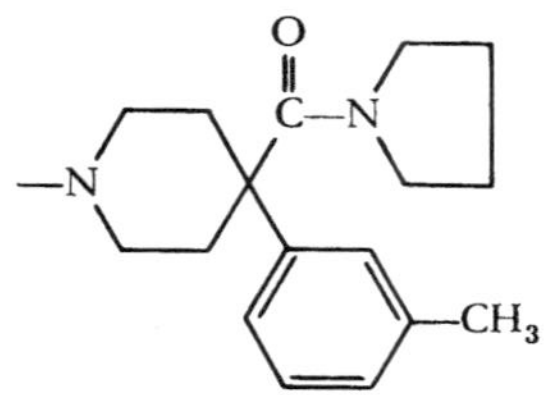	1 5 40	2.7 2.8 3.3	0.6 2.0 3.7	3.0 3.0 3.5	2.0 2.5 3.5
R-2962 Amiperone (Janssen)	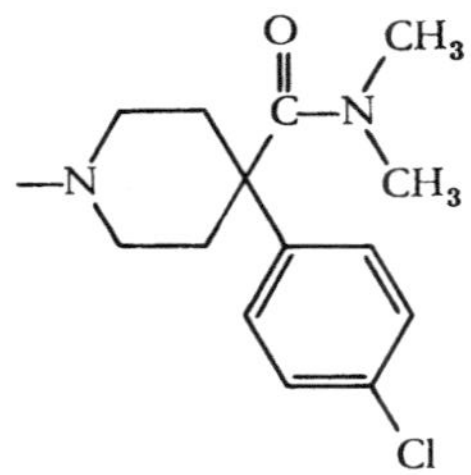	1 5 40	2.6 3.0 3.4	0.8 1.4 3.8	2.0 3.0 3.0	1.5 2.0 3.0
R-1892 Butyropipazone (Janssen)	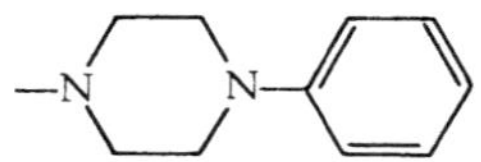	1 5 40	2.5 2.4 3.0	1.5 3.6 5.0	2.0 3.0 4.0	0.5 1.5 3.0

FR-02 (Sandoz)	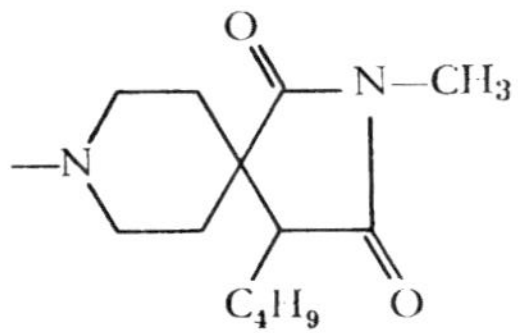	1 5 40	2.4 2.6 3.6	1.0 2.9 2.2	1.0 1.0 3.0	0 0.5 1.0
R-1647 Anisoperidone (Janssen)	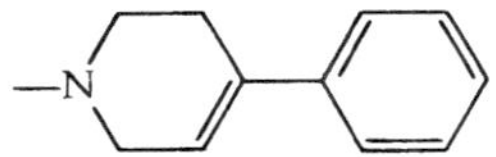 F replaced by CH_3O—	1 5 40	2.0 2.5 2.8	0.6 2.4 3.5	0 1.5 2.0	0 2.5 3.0
R-4082 Floropipeton (Janssen)	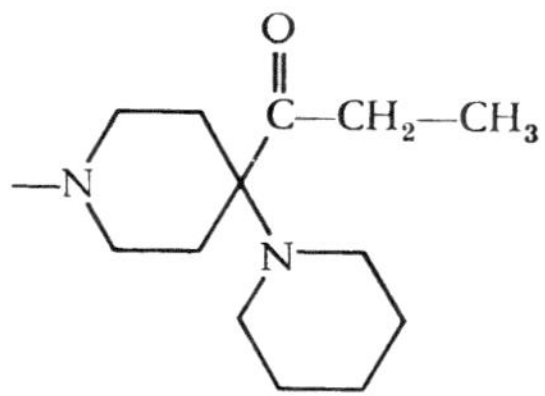	1 5 40	2.0 2.2 3.8	0.7 1.6 3.6	0 2.5 5.0	0 1.0 3.0
R-4006 (Janssen)	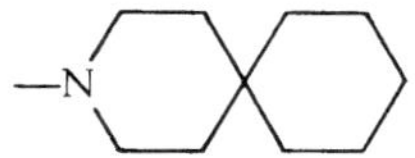	1 5 40	2.0 2.8 3.6	0.1 3.1 4.8	0 2.0 2.5	0 1.0 1.0
R-1929 Azaperone (Janssen)	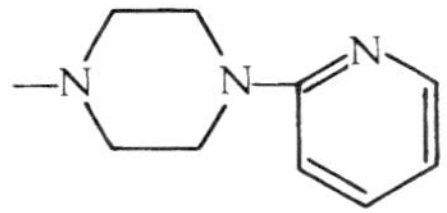	1 5 40	2.0 2.8 4.2	1.9 4.0 5.0	1.0 1.1 4.0	0 0.5 3.0

Table 5 (continued)

Code No. and generic name	Substitution of R_1	Daily dose (mg/kg)	Mammotropic index (MTI)	Body temperature drop (BTD)	Cataleptic effect (CPE)	Eyelid ptosis (PTS)
R-3264 (Janssen)	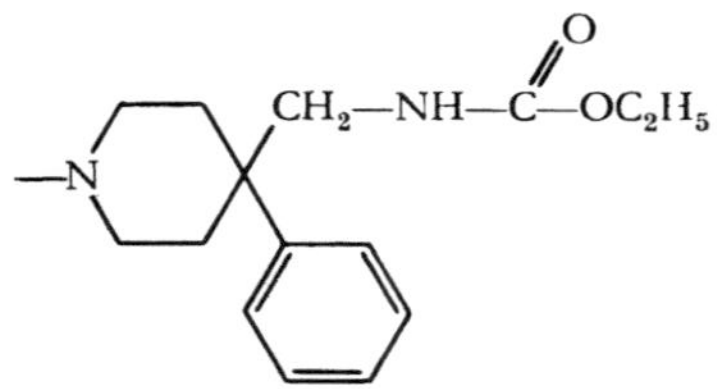	1	1.9	2.1	1.5	0.5
		5	2.9	3.8	1.0	0.5
		40	3.4	5.0	2.5	1.5
FR-33 (Sandoz) R-7158 (Janssen)	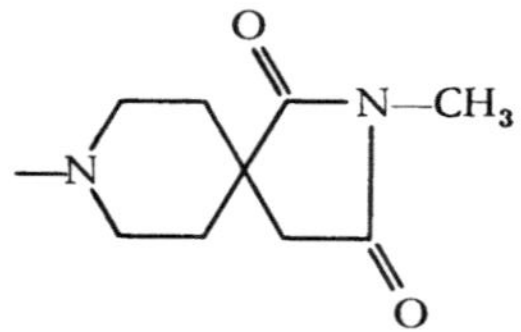	1	1.6	2.5	1.5	0
		5	1.8	3.3	2.0	0.5
		40	3.5	5.0	4.0	2.5
R-3248 Aceperone (Janssen)	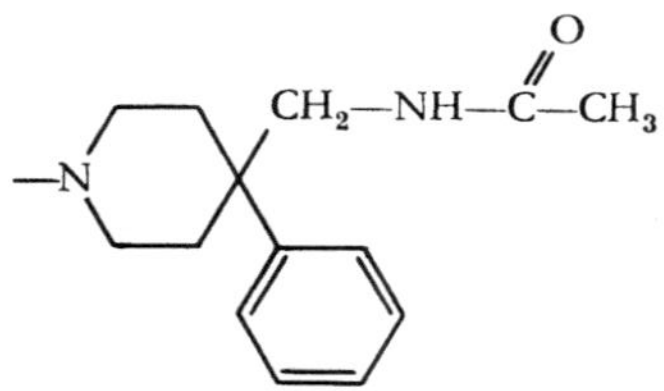	1	1.5	0.6	3.0	1.5
		5	2.3	2.0	3.0	2.5
		40	3.7	3.0	3.0	3.0
Solvent (Control)	ethanol + 0.3 N HCl $\overline{aa} \rightarrow$ 0.5 ml	0.5 ml	1.2	0	0	0

The degree of ptosis was evaluated simultaneously with the cataleptic behavior, 60 minutes after s.c. injection of the drug tested.

The 4 parameters measured and evaluated after s.c. injections into female rats of three different doses of the 23 butyrophenones are presented in Table 5. Eleven butyrophenones were found to be mammotropic at a dose of 1 mg/kg (MTI above 3.0). This result did not improve at a dosage of 5 mg/kg, but at 40 mg/kg ten more butyrophenones gave a mammotropic reaction.

All butyrophenones tested caused a significant lowering in body temperature (i. e. more than 0.6° C) even at the small dose of 1 mg/kg. MTI and BTD showed a very close correlation to each other, apparently due to a similar mechanism of action, though they differed quantitatively.

The cataleptic and ptotic reactions were also closely interrelated, though without any connection with the mammotropic effect. Ten of the butyrophenones tested proved to be cataleptic at the small dose of 1 mg/kg, seven others were cataleptic at 5 mg/kg and five became cataleptic at the highest dose of 40 mg/kg. One substance—pimozide—did not show any significant cataleptic effect within 60 minutes; its onset of action is probably very slow.

The ptosis reaction followed the pattern of the cataleptic butyrophenones except that—in addition to pimozide—clofluperol, floropipamide, FR-02 and R-4006 also showed a low degree of ptosis.

Many of these results were to be expected, as butyrophenones cause a marked hypothalamic depression, similar to the typical chlorpromazine and reserpine effects. Earlier experience has shown that the MTI and the BTD are most reliable parameters of specific hypothalamic depression, the MTI measuring the effect on the prolactin-inhibiting factor (PIF). The BTD is a standard test for hypothalamic sedation, since the thermoregulatory centers of the thalamus are connected with the emotional centers of the hypothalamus (Lissen and Parkes, 1957; Peterson and Nielson, 1964). Serotonin increases temperature, whereas noradrenaline decreases it (Feldberg, 1965). Butyrophenone derivatives, like phenothiazine derivatives, deplete both serotonin and noradrenaline and thus deprive the hypothalamus of its heat regulatory mechanism.

We concentrated mainly on the mammotropic effect of the butyrophenones, and were struck by the fact that they promote mammotropic development at lower doses than do the phenothiazines. This is probably due to their surface-spreading tendency, forming a film on biological structures, and thus becoming potent membrane permeability blockers (JANSSEN, 1967 a, b; GUTH and SPIRTES, 1964).

All neuroleptics are competitive antagonists of adrenaline, noradrenaline and dopamine, which act as neurotransmitters in the central nervous system (JANSSEN, 1967 a, b). It is postulated that haloperidol and other potent neuroleptic drugs block the dopaminergic neurones by forming a permeability-decreasing mono-layer on the membranes surrounding the cleft on the dopaminergic synapse. In this way, the dopamine molecules fail to reach the receptor site, the nerve impulses cease to be transmitted, and the animal becomes depressed. This may also be the mechanism of the BTD in the rat, since it seems that temperature is the outcome of a fine balance between the release of catecholamines and serotonin in the hypothalamus (FELDBERG, 1965), and it has been suggested that butyrophenones antagonize the influence of catecholamines, thus decreasing body temperature. The dopamine molecules are either stored in the dopamine pools (HORNYKIEWICZ, 1960) or metabolized by monoamine oxidase (MAO) in the cytoplasm. Depressing the dopaminergic neurones by butyrophenones requires much smaller doses than are needed when using chlorpromazine (JANSSEN, 1967 a, b).

The relation between the ability of the butyrophenones to antagonize catecholamines and their mammotropic effect was studied by COPPOLA, LEONARDI, LIPPMANN, PERRINE and RINGLER (1965), who suggested an indirect mechanism for prolactin release by depletion of noradrenaline from the PIF center in the median eminence. This theory was recently adopted by VAN MAANEN and SMELIK (1968), who localized the monoaminergic prolactin-inhibitory system in the tuberal part of the hypothalamus and postulated that it reacted to dopamine. The two hypotheses about the site of action of the butyrophenones may well explain the mechanisms of their mammotropic and body temperature lowering effects.

The mammotropic indices for the 23 butyrophenones tested are listed in Table 5 in descending order of their lactogenic effect at a dose of 1 mg/kg. Like other potent mammotropic neuroleptics—such as fluphenazine, trifluperazine and reserpine—the potent mammotropic butyrophenones are tertiary amines, derivatives of 4-substituted piperidine or piperazine rings. Replacement of the ketonic moiety by an aromatic radical does not seem to impair the mammotropic effect.

Small polar substituents, in para-position to the piperidine ring, increase the neuroleptic potency without, however, reducing the mammotropic effect, but lengthening of this group does impair lactogenic activity. This fact is also described in reference to the lactogenic phenothiazines (cf. Chapter 19).

The evaluation of the cataleptic effect of mammotropic drugs is obviously important, as ideal mammotropic activity should lack all concomitant cataleptic effect. The latter phenomenon, first observed by COURVOISIER, DU CROT and JULOU (1957) is caused by an activation of cholinergic processes and it is antagonized by anticholinergic and antidepressant drugs (ZETLER, 1968). It seems to be linked to the mechanism of the BTD, since hypothermia, too, can be caused by centrally acting cholinergics and may be prevented or reversed by anticholinergics (MAICKEL, STERN and BRODIE, 1963). Catalepsy is obviously related to ptosis, and is also known as "the catalepsy-ptosis observation test in rats" (JANSSEN, NIEMECEERS and SCHELLEKENS, 1965).

Summing up we feel that for the purpose of lactogenic activity a substance should have a high MTI and low BTD, CPE and PTS scores, as e. g. pimozide. The only three other substances approaching this goal among the butyrophenones tested were trifluperidol (R-2498), droperidol (R-14749) and clofluperol (R-9298).

Chapter 18

Mammotropic Effect of Miscellaneous Psychotropics Drugs

The mammotropic effect obtained with derivatives of phenothiazine (Chapter 14), tricyclic ring systems (Chapter 15), reserpine (Chapter 16), and butyrophenones (Chapter 17) prompted us to study some psychopharmaca which were assumed to act on the hypothalamus (unpublished results).

Thirty-six compounds were tested. Only four of them were found to induce mean MTI values of 3 to 3.5 at non-toxic doses: guanoclor (Vatensol, Pfizer), meclozine (Bonamine, Pfizer), methyldopa (Aldomet, Merck), tripelennamine (Pyribenzamine, Ciba). The other thirty-two compounds did not induce mammary development beyond MTI 2.8 at non-toxic doses. These inactive drugs were: amphenidone (Dornwal, withdrawn preparation), azacyclonol (Frenquel, Merrell), captodiamine (Covatin, Lundbeck), clomiphene (Mer-41, Clomid, Merrell), chlordiazepoxide (Librium, Roche), chlorothiazide (Diuril, Merck), diazepam (Valium, Roche), digitoxine (Nativelle), dyphylline (Hyphylline, Continental Labs.), fucidin (Leo Labs.), guanoxan (Envacar, Pfizer), histamine dichloride (Merck), HPt-430 (C. H. Boehringer, Ingelheim), HR-1978 (Wander, Berne), 5-hydroxytryptophane (Sandoz), hydroxyzine (Atarax, Pfizer), iproniazide (Marsilide, Roche), isocarboxazide (Marplan, Roche), Kö 592 (C. H. Boehringer, Ingelheim), lanatoside C (Cedilanid, Sandoz), melatonin (Upjohn), meprobamate (Miltown, Wallace), methamphetamine (Pervitin, Temmler), monzal (Thomae, Germany), morphine HCl and narceine HCl (Sandoz), phencarbamide (Escorpal, Bayer), D-phenylalanine (Merck), scillaren A (Transvaalin, South Africa), thalidomide (Contergan, withdrawn preparation), theophylline (Aminophylline, Searle), tybamate (Solacen, Wallace).

The cardiac glycosides which are known to induce gynecomastia in man (LEWINN, 1953) did not induce full lobulo-alveolar mammary development in the MTI test. The gynecomastia-inducing activity of digitalis preparations is attributed to their metabolic conversion into estrogen-like derivatives. Since hypothalamus-mediated mammotropism is assumed to result from protracted effects on the hypothalamus-pituitary axis, short-term administration of cardiac glycosides would not elicit mammotropic action. This was confirmed by us in animal experiments. We also had negative results with morphine and methamphetamine, which can produce lactation in highly estrogen-primed rats. Doubtless, our lower estrogen-priming allows more specific reactions, probably because of the subthreshold suppression of the PIF center. Investigations in our laboratory by BEN-DAVID (unpublished) reveal that estrogen-priming is not required for a mammotropic reaction if specific lactogenic phenothiazine compounds acting on the hypothalamus are injected into rats. In future studies of the mammotropic effects in *non*-estrogen-primed rats we hope to be able to elucidate reports claiming the non-specificity of many mammotropic effects of drugs (NICOLL, TALWALKER and MEITES, 1960; MEITES, NICOLL and TALWALKER, 1963).

The evidence presented in this and preceding chapters enables us to state that mammotropism is a quite specific property of drugs, more or less closely related to hypothalamotropic actions. This structure-activity relationship is discussed in Chapter 19.

Chapter 19

Structure-Activity Relationship of Mammotropic Drugs and Receptor Theory

The data reported in Chapters 13—18 can now be summarized to define the requirements for a mammotropic compound. Mammotropic activity is a typical side-effect of the derivatives of phenothiazine, reserpine, the butyrophenones and many hypothalamic psycho-pharmaca, which does not run parallel to the tranquilizing activity of these groups. In order to find the most mammotropic phenothiazine derivative, some hundred chemicals bearing a phenothiazine nucleus were screened in our laboratory, and the following six requirements for optimal lactogenic activity were established:

1. Changes in the phenothiazine nucleus impair the mammotropic effect, but phenothiazine sulfoxides do not lose much of their mammo-

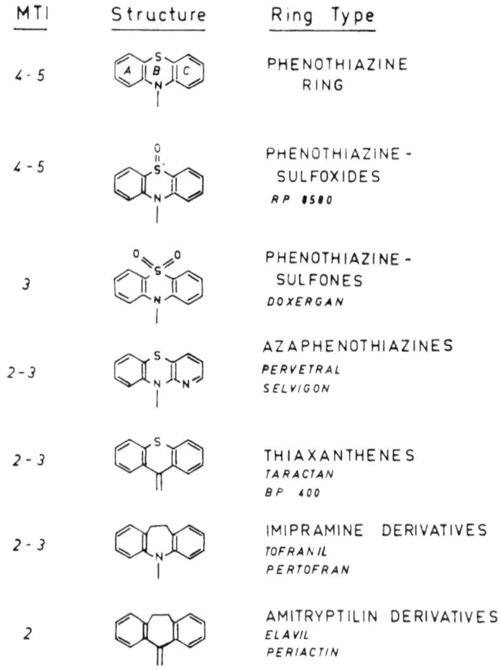

Fig. 33. Rules of mammotropic effect of phenothiazine dericatives: 1. Changes in the phenothiazine ring reduce the mammotropic effect—sulfoxides are optimal

tropic activity (Fig. 33). This indicates that the action of the S moiety is not essential for the mammotropic effect.

2. Substitution at C_2 by Cl, Br, CF_3, SCH_3, SC_2H_5, CN or $CO(CH_3)$ increases the mammotropic effect; substitution at C_2 by H or OCH_3 decreases it (Fig. 34). This indicates that a polar group at C_2 is required for the mammotropic effect.

MTI	Structure	Generic & Proprietary name
4.7	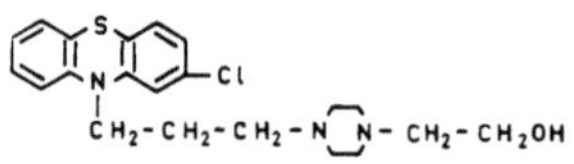	PERPHENAZINE *TRILAFON*
4.6	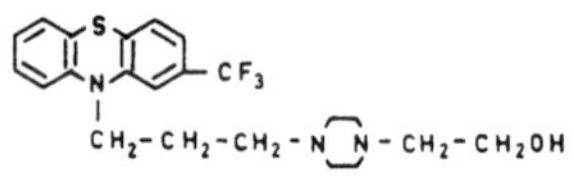	FLUPHENAZINE *PROLIXIN*
4.6	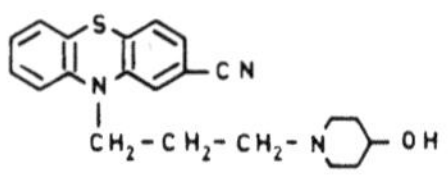	PROPERICIAZINE *NEULEPTIL*
4.2	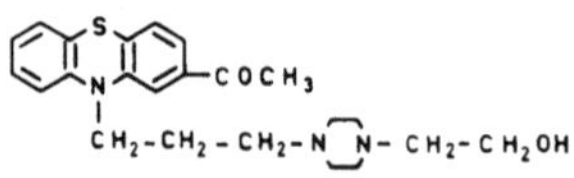	ACETOPHENAZINE *TINDAL*
3.0	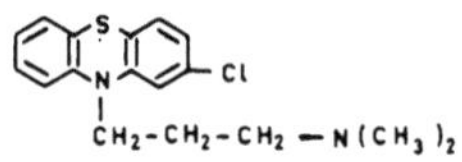	CHLORPROMAZINE *LARGACTIL*
2.8	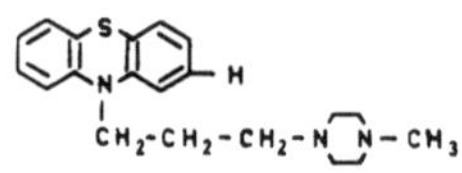	PERAZINE *TAXILAN*
2.4	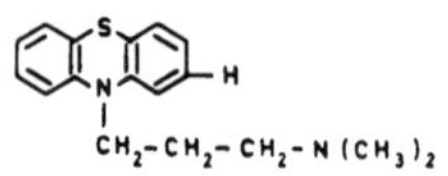	PROMAZINE *SPARINE*
1.8	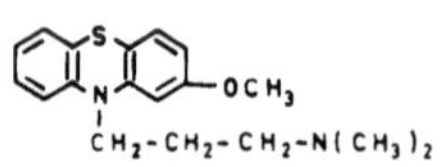	METHOPROMAZINE *MOPAZINE*

Fig. 34. Rules of mammotropic effect of phenothiazine derivatives: 2. Substitution at C_2 by Cl, Br, CF_3, SCH_3, SC_2H_5, CN or $COCH_3$ increases mammotropic effect; substitution at C_2 by H or OCH_3 decreases mammotropic effect

3. The distance between N_{10} and N_{14} must be filled by three C (Fig. 35). This allows a configuration resembling a halved hexane ring.

4. The side chain C_{11}—C_{13} must be devoid of any substitution. This points to its possible entrance into a receptor ready to receive the halved hexane ring (Fig. 36).

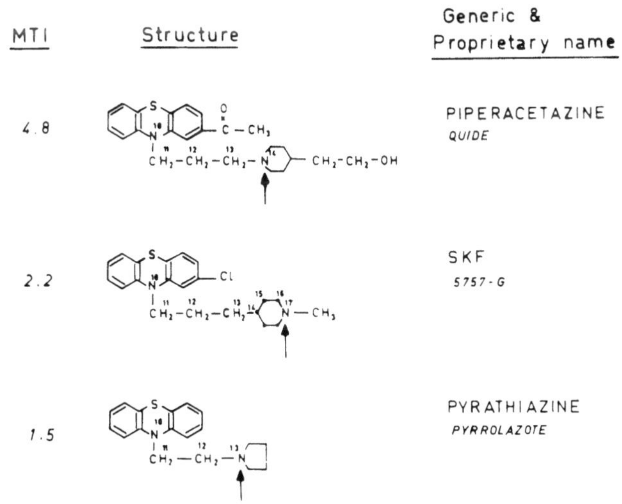

Fig. 35. Rules of mammotropic effect of phenothiazine derivatives: 3. The distance between N_{10} and N_{14} must be filled in by 3 C

5. Heterocyclic rings (e. g. piperidine, pyrrolidine or piperazine) following C_{11}—C_{13} increase the mammotropic effect (Fig. 36). This differentiates the receptor of the phenothiazine molecule from that of the steroid molecule.

6. Incorporation of C_{12} or C_{13} into the piperidine ring decreases the mammotropic effect (Fig. 37). This confirms the requirement of a halved hexane ring for the hypothalamic mammotropic receptor.

Mammotropic substances can thus be divided into two groups wich require similar but different hypothalamic receptors (Fig. 38):

A. steroids (estrogens, including stilbestrol and its derivatives);

B. cyclic compounds composed of 4—5 rings containing N (phenothiazines and their tricyclic derivatives if not prolonged by esterification, reserpine derivatives, certain butyrophenones, Fig. 39).

Both groups have the following features in common (Fig. 40):

MTI	Structure	Generic & Proprietary name
4.4	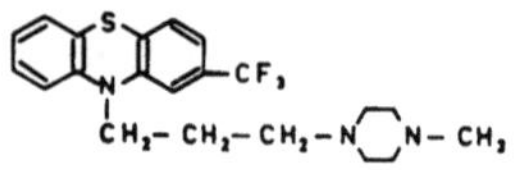	TRIFLUOPERAZINE *STELAZINE*
2.8	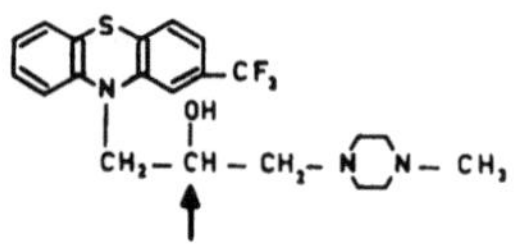	SKF *6108-J₂*
2.2	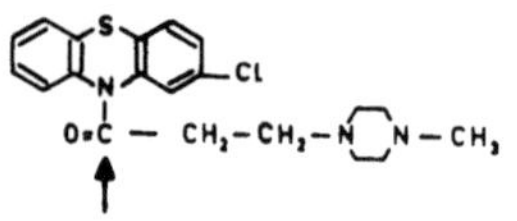	SKF *6262-A₂*
1.7	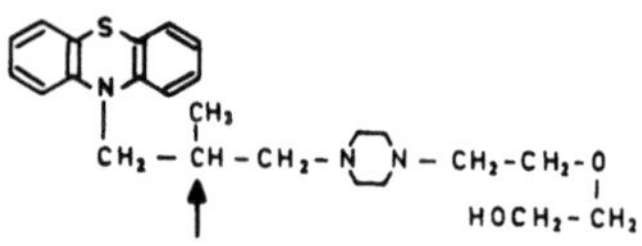	DIXYRAZINE *ESUCOS*

Fig. 36. Rules of mammotropic effect of phenothiazine derivatives: 4. An unbranched aliphatic side chain of C_{11}—C_{13}, connecting N_{10} with N_{14} is a prerequisite for mammotropic effect

MTI	Structure	Generic & Proprietary name
3	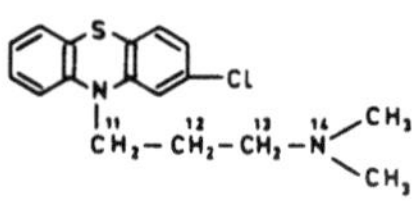	CHLORPROMAZINE *LARGACTIL* *MEGAPHEN* *THORAZINE*
4.7	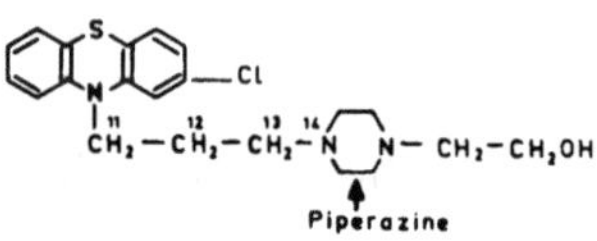	PERPHENAZINE *TRILAFON*
4.8	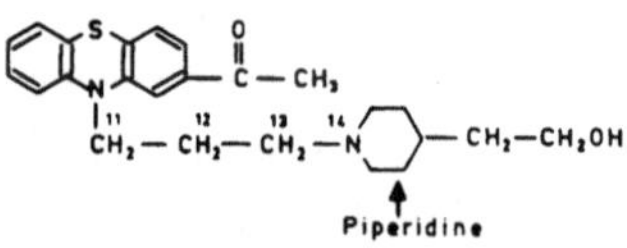	PIPERACETAZINE *QUIDE*

Fig. 37. Rules of mammotropic effect of phenothiazine derivatives: 5. A heterocyclic ring (piperidine, pyrrolidine or piperazine) following C_{11}—C_{13} increases the mammotropic effect

a) an aromatic ring A with a polar group at the same site (C_3 for steroids, C_2 for phenothiazines, C_{11} for reserpine derivatives);

b) 3—4 hydrogenated B—E rings.

The receptor for steroids can be blocked by steroids; it is related to the hypothalamic prolactin-releasing factor since clinical experience

MTI	Structure	Generic & Proprietary name
4.8	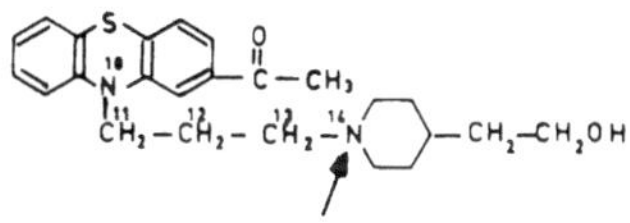	PIPERACETAZINE *QUIDE*
3.8	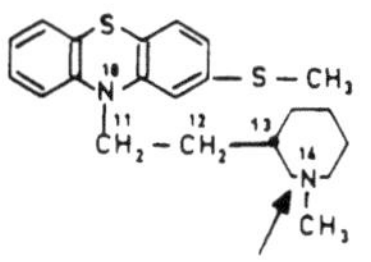	THIORIDAZINE *MELLARIL*
2.4	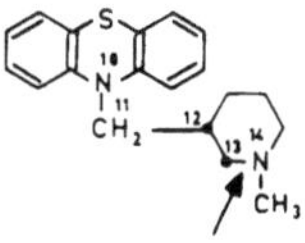	MEPAZINE *PACATAL*

Fig. 38. Rules of mammotropic effect of phenothiazine derivatives: 6. Incorporation of C_{12} or C_{13} into the piperidine ring decreases the mammotropic effect

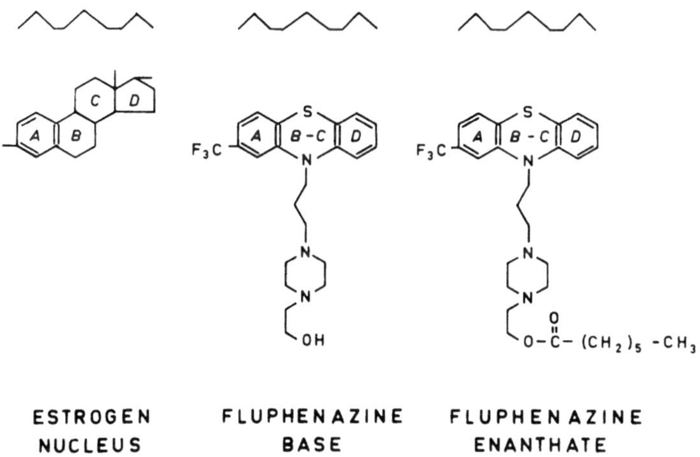

Fig. 39. Hypothetic estrogen receptor which can partly receive fluphenazine base but hardly cope with fluphenazine enanthate

has shown that it is stimulated by small doses of estrogen and suppressed by larger ones, e. g. in post-partum lactation. The receptor for the cyclic N-compounds cannot be blocked by steroids; it is linked to the hypothalamic prolactin-inhibiting factor, since clinical experience has shown that it is susceptible to psychopharmaca but not to estrogens. This experience is well known to any clinician who has tried the suppressive effect of estrogens in Chiari-Frommel Syndrome or Ahumada-del Castillo Syndrome.

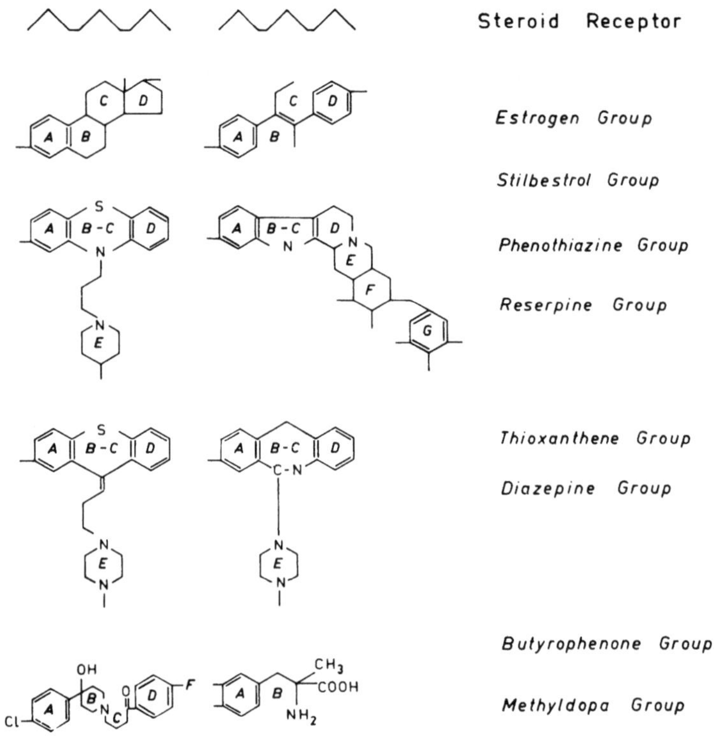

Fig. 40. Similarity of structure of lactogenic substances: 1 estrogens; 2 stilbestrol; 3 phenothiazines; 4 reserpines; 5 thioxanthenes; 6 diazepines; 7 butyrophenones; 8 methyldopa. All drugs found to be powerful prolactin releasers have the following steroid features in common: a) one aromatic ring with a substitution at C_3 ; b) at least two additional rings, or c) structural arrangements which simulate two additional rings. This structure makes them fit into the hypothalamic estrogen receptor which promotes lactation

The fact that side chains protruding from the tree-cyclic systems should consist of an unbranched 3-C-membered chain—which can fold in order to imitate half a hexane ring—seems of special importance since it proves the existence of a tripartite hypothalamic receptor sensitive to cyclic structures (Figs. 39, 40).

The validity of this conception has been accepted in pharmacology since the introduction of sympathomimetic (adrenergic) amines emanating from alkyl derivatives of ethylamine. Experience with these alkyl adrenergic amines has shown that addition of a methyl or ethyl group to the beta-C-ethylamine moiety does not yield active preparations. However, substitution by a triple-carbon (propyl) or a larger chain, which allows folding, results in preparations which pharmacologically mimic the adrenergic effects of adrenaline or noradrenaline. These new derivatives are known as alkyl or aliphatic pressor amines (1,3 dimethylamine, Forthane: 3C + 2C; N,1-dimethylhexylamine, Oenethyl: 4C + 2C; Heptamine or Tuamine: 5C + 2C; Octylamine: 6C + 2C; Nonylamine: 7C + 2C). It is obvious that an aliphatic adrenergic amine composed of less than 3C + 2C cannot fold with its head piece to fill the receptor site which normally receives the phenyl moiety of adrenaline.

These rules were confirmed by the results obtained with fluphenazine enanthate (Chapter 14). Whereas the nonesterified mother substance—fluphenazine—had high lactogenic activity (MTI 3.3—4.6), the ester when given in the same dose was inactive (MTI 1.2—2). This surprising result induced us to review the properties required of an active lactogenic drug, and we arrived at the following conclusions:

a) The enanthate esterification deprives fluphenazine of its lactogenic activity because it adds bulk to the five-ring system which prevents it from entering a hypothetic hypothalamic receptor. It, therefore, needs to be split up at its site of injection in order to reach the hypothalamus (Fig. 39).

b) The hypothalamic receptor must have at least three main sites for natural ring systems which stimulate lactation in women and animals, i. e. the steroids (Fig. 39).

c) Most substances described as potent lactogenic stimulants in this report have a ring formation which imitates steroid rings (Fig. 40).

d) The six rules for active phenothiazine derivatives (Figs. 33—38) fit extremely well into this theory. If they are redrawn, Fig. 40 will present the conclusion that at a typical position of an aromatic ring, which is identical for all lactogenic substances (estrogen: C_3, stilbestrol: C_3, phenothiazine: C_2, reserpine: C_{11}), there must be a substitution which can act as a polar group within the body.

e) This rule is supported by the puzzling fact (Chapter 16) that reserpine with a methoxy group at C_{11} is lactogenic, whereas its isomer methoserpidine—with the methoxy group at C_{10}—is inactive as a lactogen, but as effective a tranquilizer as reserpine itself.

f) Fig. 40 also furnishes and explanation why thioxanthene derivatives, such as P-4657-B, and the Wander compounds, such as SUM 3170 and HF 2159 (Chapter 15), are lactogenic. Even haloperidol and its derivatives (Chapter 17) could be conceived in such a way that they fit into the hypothalamic steroid acceptor, as might methyldopa (Chapter 18).

Thus we arrive at the conclusion that study of the structure-activity relationship in this research furnishes us with a simple rule for future synthesis of lactogenic substances. These should have at least 3 rings—the first one aromatic—to fit the hypothalamic lactation receptor. The aromatic ring should have a polar group or a substitution at C_3 to resemble the steroids. It should have one or two piperidine, piperazine, pyrrolidine or even azepine rings attached. Side chains should be avoided, and up to 6 rings in all could be expected to be active lactogens. Synthesis of ring compounds to fit these rules and to obtain highly lactogenic substances which would specifically affect the LTH centers of the hypothalamus, but exert no tranquilizing or sedative effect, is now being planned in our laboratory.

Another aspect of the above receptor hypothesis is the proof recently furnished by us that estradiol and phenothiazine derivatives compete for the same brain receptor (MISHKINSKY, GIVANT, EYLATH and SULMAN, unpublished data). Labeled and unlabeled phenothiazine derivatives (0.05—50 mg/kg) were given by intravenous injection to ovariectomized rats and their cerebral uptake studied. When tritiated estradiol (50 c/mM) was injected i.v. (1 μg/kg) 30 seconds after the phenothiazines, its uptake in the anterior pituitary and median eminence depended upon the quantity of phenothiazines given before. The dose-response curve revealed only 10—50% of estradiol uptake after administration of perphenazine, aminopromazine or phenothiazine. This proves that phenothiazine derivatives compete quantitatively with estradiol for a common steroidal receptor in the hypothalamus and anterior pituitary. Control measurements with other brain sectors did not show any significant uptake of steroids or of phenothiazine derivatives (Fig. 41). The reverse experiment, giving estradiol before the phenothiazine, did not reveal any significant competition.

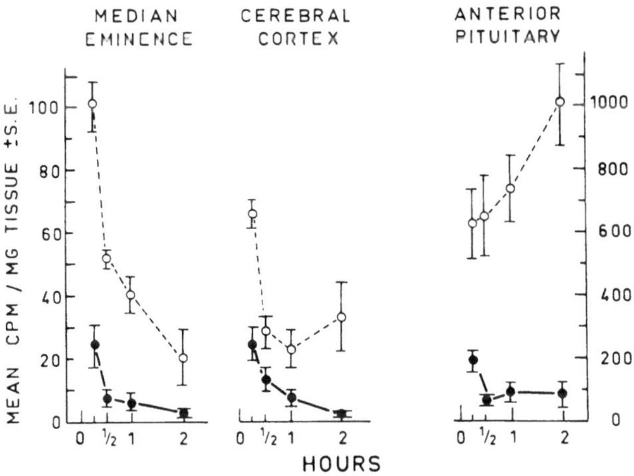

Fig. 41. Block of the estradiol receptor by perphenazine given s. c. 30 sec before i. v. injection of estradiol. The rats were killed 1/4, 1/2, 1 and 2 hours after estradiol injection. Median eminence block: 70—90%; Anterior pituitary block: 70—90%; Cerebral cortex block: 30—45% only. Note the specific uptake and block of both substances in median eminence and pituitary which has its optimum at 1/4—1/2 hr. but lasts beyond 2 hrs. Pituitary uptake of estradiol is, as a rule, 10 times higher than that of median eminence. ●——● Estradiol after Perphenazine; ○- - - -○ Estradiol alone

Chapter 20

Pituitary-Ovary Axis in Hypothalamic Lactation

Perphenazine was found to be an effective agent for elucidation of the mechanism of hypothalamic lactation (Ben-David, Dikstein and Sulman, 1965). It stimulates pituitary prolactin secretion *in vitro* in hypothalamus-pituitary co-culture (Danon, Dikstein and Sulman, 1963) and, *in vivo* via the synthesis and release of pituitary prolactin in intact animals (Ben-David and Chrambach, 1970). Similarly, perphenazine (Velardo, 1958), reserpine and chlorpromazine (Barraclough, 1957) have also been reported to induce pseudopregnancy, indicating stimulation of the secretion of luteotropic hormone which produces progesterone (VELARDO, DAWSON, OLSEN and HISAW, 1953). Estrogen, too, stimulates prolactin secretion (RAMIREZ and MCCANN, 1964). Thus, all ovarian hormones may play a role in mammary gland growth. LYONS showed already in 1958 that, in the hypophysectomized rat, ovarian and hypophysial hormones are required for normal mammogenesis. Only later did TALWALKER and MEITES (1961) demonstrate that, in the hypophysectomized-ovariectomized rat anterior pituitary hormones are mammogenic by themselves if given in sufficient quantity. It can, thus, not be determined whether perphenazine treatment induces lactation via the hypothalamus and the anterior pituitary—by stimulating the release of pituitary mammotropic hormones (LTH and STH)—or also via the hypothalamus, the anterior pituitary and the ovaries, i. e. by stimulating ovarian hormone secretion as well. In order to elucidate the role of the ovaries in perphenazine-induced lactation, experiments were carried out (BEN-DAVID, 1968 b) in intact and ovariectomized, estradiol-primed and non-primed adult rats and in juvenile rats, in which simultaneous determination of the DNA content of the mammary glands and histological studies of the mammary glands, ovaries, uterus, and vaginal smears were performed after 5 days' treatment with perphenazine. In bilaterally ovariectomized rats treatment was started 7 days after operation. For priming, 8 μg estradiol was injected for 10 days prior to perphenazine treatment.

In normal (non-primed) adult rats, full mammary gland development (MTI : 5) was obtained by 5 daily injections of 10 mg perphenazine/kg. In ovariectomized adult rats the mammotropic effect of perphenazine was always lower than in the intact rats, the MTI not exceeding 3.8. This lesser response was, however, overcome by first priming the ovariectomized animals with small doses (8 μg/d) of estradiol for 10 days. In estradiol-primed rats a daily dose of 10 mg perphenazine/kg produced a maximal lactogenic response (MTI : 5) in both intact and ovariectomized animals.

The DNA content of the mammary glands increased with the perphenazine dosage, but not RNA. A sharp increase in both DNA/mg tissue and total DNA content was obtained when the perphenazine dose was increased to 5 mg/kg/day. However, the DNA/mg tissue rose less sharply than total DNA content when higher doses of perphenazine were employed. In juvenile rats, perphenazine treatment alone (10 mg/kg/day, for 5 days) failed to induce significant mammary growth (MTI: 2) even though the absolute weight of the pituitary and that per 100 g body weight were substantially ($P < 0.05$) increased by 20% and by 29% respectively. However, in estradiol-primed juvenile rats, similar perphenazine treatment induced considerable mammary growth (MTI: 4.8). Perphenazine administered to juvenile rats caused no significant change in the weight of ovaries or uterus, nor did it induce precocious opening of the vagina.

The changes in the mammary glands, ovaries, and uterus of intact adult non-primed rats receiving perphenazine for 5 days, starting at early diestrus, are illustrated in Fig. 42. The mammary glands were found to be highly developed and the lactogenic response increased with the dose. Milk was present in the mammary glands of all animals receiving the excessive dose of 15 mg/kg. Examination of the ovaries showed a greater number of corpora lutea (3—6) in the ovaries of treated animals than in the ovaries of the control rats (0—2), an observation confirmed by sectioning the ovaries.

No animal among the perphenazine-treated rats had a fluid distended uterus (which usually indicates estrogenic activity), as against 40% of the control animals. Histological investigation of the uteri of treated rats (Fig. 42) showed that basal vacuoles ("lucid zones") appeared in the endometrium which contained many more nuclei than that of the control. Moreover, diestrus persisted in the treated rats until about 7 days after the end of treatment.

Histological studies and determination of the DNA content of the mammary glands after administration of various doses of perphenazine showed that the degree of mammary growth was directly proportional to the doses employed. While total DNA content of the left inguinal mammary pad showed a continuous sharp increase parallel with the increase in perphenazine dosage up to 15 mg/kg, only a limited rise in DNA per 100 mg mammary tissue was obtained with doses higher than 5 mg/kg, indicating that, on reaching their maximum degree of development, the mammary cells started to produce milk rather than to divide. At the same time, however, larger doses of the drug caused further growth of mammary tissue, as indicated by the increase in total DNA.

In estradiol-primed rats a comparison between the lactogenic effect of perphenazine in intact and ovariectomized rats showed no difference in the mammary response at any of the doses given, while in non-primed rats there appeared to be greater mammary response in the intact animals. Moreover, in the non-primed ovariectomized rats the maximal mammary response did not exceed MTI 3.8, even when a higher dose (10 mg/kg/day) was employed. It may be assumed, therefore, that, in order to achieve full lactogenic response after perphenazine treatment, the mammary tissue must be primed with small

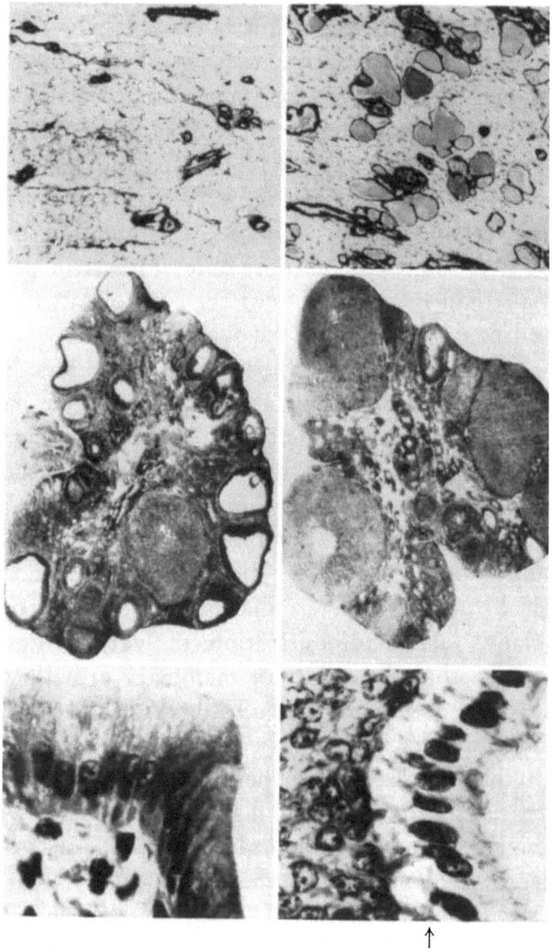

Fig. 42. Effect of 5 days' treatment with 5 mg perphenazine/kg/day on the mammary glands, ovaries and uteri of rats weighing 95—105 g. Treatment was started at early diestrus: l: control; r: treated rat. Note mammary development, luteotropic effect on the ovary and "lucid zone" (arrow) in treated uterus

amounts of estrogen, either by physiological estrogen present in the intact animal, or by exogenous estradiol in ovariectomized rats. Estrogen seemingly acts as a permissibe factor in perphenazine-induced lactation.

Partial mammary growth in ovariectomized rats induced by perphenazine (MTI : 3.8) seems to result mainly from stimulation of the secretion of mammotropic hormones from the anterior pituitary. This assumption is supported by the results of TALWALKER and MEITES (1961) who showed that anterior pituitary hormones alone, if given in sufficient amounts, induce mammary growth in the ovariectomized-hypophysectomized rat.

That the presence of the ovaries is not essential for perphenazine to exercise its full mammogenic effects is evident from the fact that a dose of 10 mg/kg/day for 5 days was fully effective in ovariectomized rats provided they had been primed with estradiol. Moreover, full lactation has also been obtained in adult non-primed male rats treated with perphenazine (BEN-DAVID, 1968 b). The state of the ovaries provided no evidence that the mammotropic effect of perphenazine was due to release of FSH evoking an increase in estrogen production by the ovaries. ECKSTEIN (1962) had reported that the ovaries of juvenile rats were very sensitive to exogenous FSH and that their weight greatly increased after administration of PMS. In the study by BEN-DAVID (1968 b) perphenazine treatment of juvenile rats caused no increase in ovarian weight, in neither estradiol-primed rats nor in non-primed juvenile rats. Similarly, no increase in uterine weight was found, indicating that FSH secretion was not stimulated by perphenazine. In fact, it has been reported that phenothiazine derivatives, e. g. chlorpromazine and also reserpine, block rather than stimulate FSH-LH secretion (SULMAN and WINNIK, 1956; WINNIK and SULMAN, 1956; NAGATA, 1957; BARRACLOUGH and SAWYER, 1957). BARRACLOUGH (1957) has also reported the induction of pseudopregnancy in rats by chlorpromazine and reserpine (cf. als KHAZAN, SULMAN and WINNIK, 1960).

When perphenazine was given at an appropriate stage of the cycle (starting at early diestrus), a typical "lucid zone" was obtained in the endometrium (Fig. 42). The "lucid zone" was first described in rats by VOKAER in 1950 as indicating the existence of active corpora lutea (pseudopregnancy). However, it is well recognized that, in rats and mice, the corpora lutea are not truly active during the normal estrous cycle, and a state of pseudopregnancy can be achieved only after release of adenohypophysial luteotropin which changes the cycling corpora lutea into active endocrine organs secreting mainly progestational compounds (VELARDO, DAWSON, OLSEN and HISAW, 1953). VELARDO (1958) also reported that a pseudopregnant state in rats could be achieved by perphenazine administration, and he confirmed the existence of this state by the induction of deciduomata in the rats.

We confirmed VELARDO's finding by reproducing the occurrence of the "lucid zone" in the endometrium of perphenazine-treated rats (Fig. 42). These experiments show that in virgin adult female rats mammary gland development and milk secretion were induced by 5 days' perphenazine (Trilafon) treatment. The drug, however, failed to induce similar lactation in either ovariectomized adult animals or in juvenile rats unless the animals were pretreated with small quantities of estradiol, indicating that, when the ovaries are absent or almost inactive, as in juvenile rats, estradiol-priming is a prerequisite for perphenazine-induced lactation. When perphenazine was given to adult rats starting at early diestrus, the following changes occurred: 1) a continuous stage of diestrus which lasted until about 7 days after the end of treatment; 2) the ovaries were predominantly occupied by corpora lutea; and 3) the "lucid zone", typical for pseudopregnancy, appeared in the uterine endometrium.

It is concluded that perphenazine treatment stimulates endogenous secretion of LTH, which in rats is identical with prolactin, and thus promotes lactogenesis.

Chapter 21

Pituitary-Adrenal Axis in Hypothalamic Lactation

The role of the adrenocortical hormones in mammary differentiation has not yet been fully investigated. Some results indicate, however, that these hormones influence ductal as well as alveolar development. It has, for instance, been reported that desoxycorticosterone stimulates ductal growth in the ovariectomized (SMITH and BRAVERMAN, 1953), and that injections of cortisol can enhance alveolar development in intact female rats (JOHNSON and MEITES, 1955). These observations confirm the widely held concept that ovarian steroids (LYONS, LI and JOHNSON, 1958) and adrenocorticosteroids (JACOB-

SOHN, 1962 a, b) are essential but not indispensable for mammary lobulo-alveolar growth in rats. There remains the decisive fact that lobulo-alveolar development is possible in ovariectomized-adrenalectomized-hypophysectomized rats injected with large doses of prolactin and somatotropin (TALWALKER and MEITES, 1961; MEITES, 1965).

While ovarian steroids seem to act as permissive factors in hypothalamic lactation (BEN-DAVID, 1968 b), the role of the adrenal corticosteroids in perphenazine-induced lactation is still unknown. This question is of special importance because of the above controversial reports concerning the role of adrenal corticosteroids in normal growth and secretion of the mammary gland. Whe have therefore undertaken a series of experiments in adrenalectomized, ovariectomized and adrenalectomized-ovariectomized adult female rats in order to elucidate the role of the pertinent steroids in perphenazine-induced hypothalamic lactation (BEN-DAVID and CHRAMBACH, 1969).

Intact, adrenalectomized, ovariectomized and adrenalectomized-ovariectomized adult female rats received 5 daily injections of 5 mg/kg/d perphenazine or control solution. All operations were performed under ether anesthesia on the first day of treatment prior to the first injection. Normal saline and 5% glucose solutions were given freely to all animals, which were kept at 24° C. The animals were killed on the sixth day. Five days after adrenalectomy alone had been performed, the mammotropic index was high (MTI: 3.5) and the mammary glands underwent no involution, whereas the mammary glands of intact control animals had an MTI of 2.4 only. Removal of both ovaries and adrenals resulted in involution of the mammary glands (MTI: 1.9). On the other hand, perphenazine treatment, which stimulated endogenous secretion of prolactin, resulted not only in prevention of mammary involution 5 days after adrenalectomy and ovariectomy, but also in mammary growth (MTI: 3.5); it also brought about an identical degree of mammary development (MTI: 3.5) in adrenalectomized rats. Still, in adrenalectomized or adrenalectomized-ovariectomized rats, the mammotropic effect of 5 mg/kg/d perphenazine was less than that obtained in intact rats (MTI: 4). It, thus, seems that lack of adrenocortical steroids in female rats has no effect on the degree of mammary growth in untreated animals, while in perphenazine-treated animals it slightly impairs the degree of mammary development. This is best shown by measuring the milk yield, as described in Fig. 43.

Absence of both ovarian and adrenocortical steroids resulted in involution of the mammary tissue of the untreated rats as well as impairment of the mammotropic effect of perphenazine. It seems most likely that, in the rat, the presence of minimal amounts of adrenocortical hormones is necessary for full development of the mammary

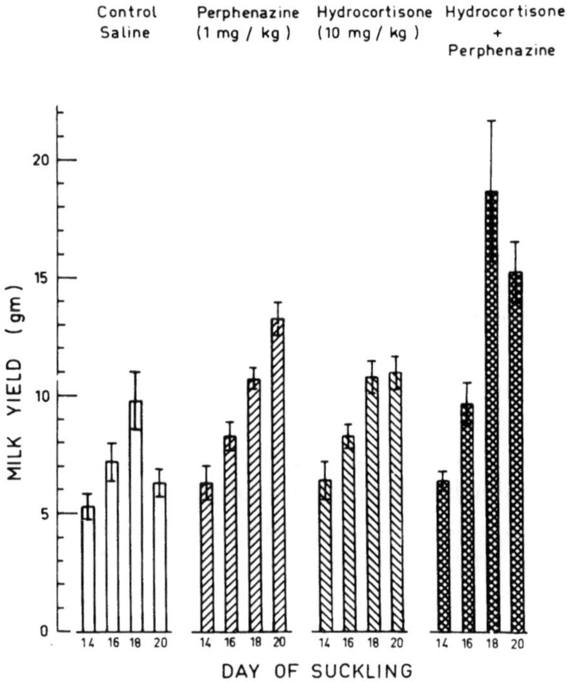

Fig. 43. Milk yield in nursing rat mothers: Hydrocortisone improves effect of perphenazine

tissue and initiation of milk secretion. In other words, the adrenal cortex is less important and possibly not essential at all for the first stages of mammary growth, i. e. mammogenesis, as evidenced by development of the ductal and lobulo-alveolar systems, while the cortical steroids seem to be necessary for the promotion of advanced development of the lobulo-alveolar system (colostrogenesis) and milk secretion (lactogenesis). Similar results were obtained in other experiments by Nicoll and Meites (1964), Hamberger and Ahrén (1964) and Meites (1965).

These results were further confirmed when a corticoid was added to the above set-up. In adrenalectomized or in adrenalectomized-ovariectomized rats, the addition of hydrocortisone to perphenazine treatment resulted in maximal development of the mammary tissue with copious milk secretion (MTI: 5). Thus, although prolactin is a powerful stimulus to milk secretion, its maximal effect cannot be achieved unless glucocorticoids are present. While hydrocortisone alone does not produce a secretory response (BARNAWELL, 1967), it potentiates the galactogenic property of prolactin.

Results of the present study might explain the many discrepancies in literature concerning the role of the adrenocortical hormones in lactation. It seems that, in any discussion of this problem, one must differentiate between the primary process of the actual mammary growth (mammogenesis and colostrogenesis) and the process of conversion of the developed gland into secreting structures (lactogenesis and galactopoiesis). With regard to the former, we found that, in the absence of steroids, the mammary glands underwent considerable growth in response to endogenous prolactin produced by perphenazine, but they never approached maximal levels. These results confirm the *in vivo* observation of MEITES (1965) that mammary growth could be achieved in male rats in the absence of gonadal and adrenocortical hormones by treatment with prolactin and growth hormone. The *in vitro* studies reported recently by DILLEY and NANDI (1968) also showed that mammary growth could be maintained in the absence of steroids. On the other hand, our findings in the rat are also in agreement with those reported by NELSON, GAUNT and SCHWEIZER (1943), FOLLEY (1952 a), and by MEITES and NICOLL (1966) who claimed that adrenocortical hormones were important for the second stage of milk secretion. Different species, however, have different requirements for milk secretion. While both mouse and rat require LTH and/or STH plus corticoids as a minimum, the rabbit may need only either one (MEITES, HOPKINS and TALWALKER, 1963).

Summarizing we must bear in mind that *in vivo* adrenocortical hormones seem to be essential for milk secretion but that they are not necessary for the first stages of mammary growth. This holds true also in perphenazine-induced hypothalamic lactation—at least for the rat; other species may have different requirements. The requirements *in vitro* are described in Chapter 6.

Chapter 22

Pituitary-Thyroid Axis in Hypothalamic Lactation

a) Role of Thyroid in vivo and in vitro

A deficiency in thyroid hormone may result in reduced mammary growth, although the rat appears to be an exception in this respect (KRAGT and MEITES, 1965). Since hypothyroidism has been found to be associated with a decrease in secretion of pituitary LTH (MCQUEEN-WILLIAMS, 1935; MEITES and TURNER, 1947 a), of STH, and ACTH (CONTOPOULOS, SIMPSON and KONEFF, 1958) as well as with irregularities in FSH-LH secretion (LEATHEM, 1959), the retardation of mammary development may be partly secondary to changes in secretion of these hormones. There is evidence that thyroid hormones can stimulate LTH and perhaps STH secretion by the rat anterior pituitary.

The *in vitro* ability of thyroxine and triiodothyronine to stimulate pituitary prolactin release is in agreement with *in vivo* experiments in which thyroid hormones were reported to stimulate pituitary prolactin secretion and milk production (GROSVENOR, 1961; MEITES, 1959 a, 1963). Thyroidectomy (MCQUEEN-WILLIAMS, 1935) or thiouracil feeding (MEITES and TURNER, 1947 a) result in diminished pituitary prolactin content in rats. Direct stimulation of pituitary prolactin release by thyroid hormones *in vitro* does not exclude the possibility that *in vivo* these hormones may also promote prolactin release through the hypothalamus.

Thyroxine alone, and in combination with other hormones, has been shown transiently to increase milk yield in lactating rats (GROSVENOR and TURNER, 1959 a, b; DJOJOSOEBAGIO and TURNER, 1964) and in cows (BARTLETT, BURT, FOLLEY and ROWLAND, 1954), and to cause mammary gland development in mice (ANDERSON and TURNER, 1963). In the goat, removal of the thyroid results in a fall in milk yield of various degrees (SMITH, STOTT and WALKER, 1957). That is not due to the absence of the thyroid becomes evident from the fact that lactation is not affected when the thyroid is destroyed by radioiodine, leaving the parathyroid essentially normal (BRUCE and SLOVITER, 1957). These experiments suggest that the thyroid is not essential for milk secretion. It seems, therefore, preferable to approach the problem by avoiding operations and using "pharmacologically thyroidectomized animals", which can be achieved by thiouracil treatment.

b) Thyroid and Hypothalamic Mammary Growth

As it was not known whether or not the thyroid hormones were involved in perphenazine-induced hypothalamic lactation, this problem was studied by assessing the effect of propylthiouracil (PTU) and triiodothyronine (T-3) on physiological as well as on hypothalamic lactation (BEN-DAVID, DIKSTEIN and SULMAN, 1965). The mammotropic effect was evaluated by the mammotropic index (MTI). In the case of propylthiouracil treatment, 0.3% of the powder was added to the diet and given for 2 weeks prior to, and during, the 5 days' treatment with perphenazine. For estimation of milk yield, lactating rats, each nursing a litter of 6, were used. Beginning on the 7th day post-partum, daily s.c. injections of different drugs were given until the 20th day. Daily weights of dams and litters were recorded from the 7th to the 20th day post-partum and milk yield was estimated from the litters' weight gain during one hour of nursing. Ten hours before suckling, litters were isolated from their mothers without any food supply. One minute prior to suckling, 1 IU of synthetic oxytocin (Syntocinon) was given to all dams to obtain complete milk ejection.

A dose of 5 mg/kg/d perphenazine for 5 days caused considerable mammary gland development in rats primed with subthreshold doses of estradiol (MTI : 4.7), while triiodothyronine sodium (T-3) in a dose of 0.03 mg/kg produced no better mammotropic effect than saline (MTI : 1.2). Moreover, no dose of T-3 (0.003—0.03—0.3 mg/kg) significantly improved the mammotropic effect achieved by perphenazine (MTI : 4.2, 4.4, 4.6). On the other hand, propylthiouracil, which inhibits endogenous thyroxine production, caused mammary gland development in estrogen-primed rats when given together with saline and even improved the mammotropic effect of perphenazine (MTI : 5). This mammotropic effect obtained by propylthiouracil was abolished by additional treatment with 0.03 mg/kg/d T-3 in both experiments: propylthiouracil alone yielded MTI : 1.0, and propylthiouracil + perphenazine MTI : 4.4.

These results clearly indicate that a difference exists between the roles of endogenous and exogenous thyroxine in mammary development. It was shown that in perphenazine-lactating rats the mammotropic effect of propylthiouracil is due to a decrease in endogenous circulating thyroid hormones, since the addition of exogenous T-3 prevented mammary gland development altogether.

c) Thyroid and post-partum Galactopoiesis

A low dose of perphenazine (1 mg/kg/d), given to lactating rats, slightly increased milk yield during the whole period of suckling, and a sharp increase was obtained on the 20th day when the drug significantly prevented the natural decrease in milk yield observed in the control animals (Fig. 44). Propylthiouracil (which suppresses endo-

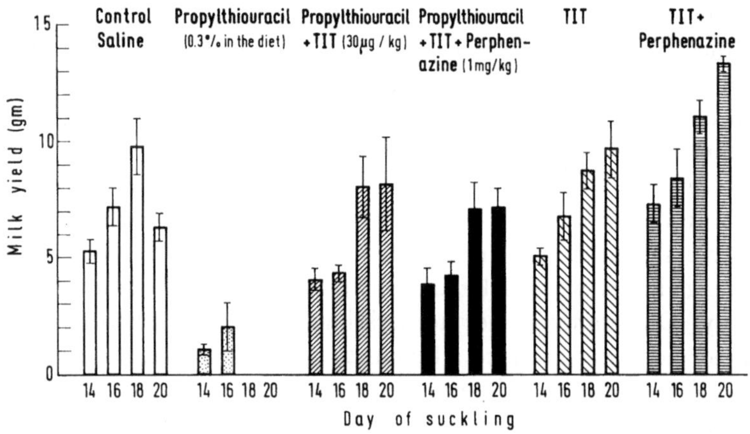

Fig. 44. Milk yield in nursing rat mothers. Propylthiouracil depresses milk yield; perphenazine and triiodothyronine (TIT) increases it

genous production of thyroxine) strongly suppressed milk production, so that the mothers failed to secrete milk on the 18th day post-partum, and the litters died on the 20th day. This suppression was prevented when combined treatment with T-3 (0.03 mg/kg/d) and propylthiouracil (0.3% in the diet) was given. Hence, T-3 did not significantly alter the milk yield either of normal lactating rats or of perphenazine-treated lactating animals. It prevented, however, the suppression of milk yield obtained by propylthiouracil.

d) Conclusion

While, physiologically, thyroid hormones inhibit mammary gland development in the first stage of lactation, they are essential for milk production in the second or third stage. On the other hand, exogenous T-3 did not show significant mammotropic effect, either on mammogenesis or on colostrogenesis, and a slight interference could even be observed. The mechanism by which thyroid hormones inhibit the first stages of mammary gland development is not clear. However, other authors (LEONARD and REECE, 1941; TRENTIN, HURST and TURNER, 1948; MEITES and KRAGT, 1964) also observed a depression of mammary gland development after administration of thyroid hormones. Propylthiouracil which stimulated mammary gland development and inhibited milk secretion, by decreasing circulating thyroid hormones, was active in artificial perphenazine-induced hypothalamic lactation as well as in normal post-partum lactation. This fact again indicates that thyroid hormones inhibit mammogenesis and colostrogenesis, but promote lactogenesis and galactopoiesis.

Chapter 23

Prolactin Release by Hypothalamus of Nursing Rat Mothers

It is well known that the hypothalamus governs the synthesis and release of tropins from the pituitary gland in different ways: hypothalamic stimulation is needed for release of most hypophysiotropins, with the exception of prolactin which is inhibited by the hypothalamus. The prolactin-inhibiting activity of hypothalami taken from cycling animals was proved and confirmed in many species (MEITES, NICOLL and TALWALKER, 1963), but, as prolactin secretion was believed to have no peripheral feedback mechanism, many workers looked for a competitive or complementary mechanism to the inhibitory one exerted by the hypothalamus.

The amount of milk produced by lactating rats increases steadily throughout lactation (MAYER, 1935; COX and MUELLER, 1937). GROSVENOR and TURNER (1958 a) found that pituitary prolactin levels in lactating rats increased from the second to the sixth day post-partum, while REECE and TURNER (1937) reported a 2-fold increase in anterior pituitary lactogen content in rats following parturition. WELLER showed in 1963 that hypothalami from milking cows had a high mammotropic effect when injected into estrogen-primed rats. The view that rat hypothalamic tissue can induce mammary secretion in rats was advanced by MEITES, NICOLL and TALWALKER (1963). RATNER and MEITES (1964) found that extracts from hypothalami of post-partum rats yielded *in vitro* more prolactin when incubated with normal pituitary glands than extracts prepared from hypothalami of virgin rats. KRAGT and MEITES reported in 1965 that the hypothalamus of a parent pigeon, on the day the young are hatched, stimulates prolactin release by the pigeon pituitary gland *in vitro*.

Indications that the CNA exerts a stimulating effect on prolactin secretion came from observations in humans and in farm animals. MEITES, TALWALKER and NICOLL (1960 a) were the first to report that injections of rat hypothalamic extracts initiated mammary secretion in estrogen-primed rats. In 1962, TURNER and GRIFFITH postulated that, under normal conditions, a "lactogen-releasing factor" builds up in the hypothalamus and lactogen stores in the anterior pituitary gland, and that, following a suckling stimulus (oxytocin release), the hypothalamic factor "is discharged into its portal blood system and stimulates the discharge of lactogen from the anterior pituitary." They also found that the low prolactin levels in the anterior pituitary gland of most species during pregnancy increase markedly following parturition (TURNER and GRIFFITH, 1962). These findings have been confirmed by us and recently also in *in vitro* work by our group (unpublished results). Thus, it seems that the prolactin-releasing activity of the post-partum rat hypothalamus is enhanced, possibly due to decreased production of the prolactin-inhibiting factor and increased production of a prolactin-releasing factor. Inasmuch as the hypothalamic homogenates contain numerous mediators, e. g., serotonin, catecholamines, acetylcholine and histamine, the possibility must be considered that these agents influence the *in vivo* release of prolactin in the test rats. It is generally believed (MEITES and NICOLL, 1966) that the effectiveness of a hypothalamic extract *in vivo* is due to its content of substances which can stimulate prolactin release indirectly, and MEITES, NICOLL and TALWALKER (1963) suggested that hypothalamic tissue induces adrenocortical stimulation which in turn elicits release of prolactin from the pituitary gland.

MISHKINSKY, KHAZEN and SULMAN (1968) studied the problem in adult virgin rats which had been primed with estradiol 8 μg/rat/day for 10 days and injected for a further 5 days with different hypothalamic and pituitarian extracts. The homogenates of hypothalami and of pituitaries of the donor rats were prepared in 3 ml of saline with a cone-glass electric homogenizer. The mother rats were sacrificed by a blow on the head on the specified day of lactation, and their hypothalami and pituitaries were removed within 1 min of death; hypothalami and pituitaries of virgin rats of corresponding ages and weights served as controls. Only fresh homogenates were used, in a proportion of 0.6 hypothalamus or pituitary gland per experimental rat per day.

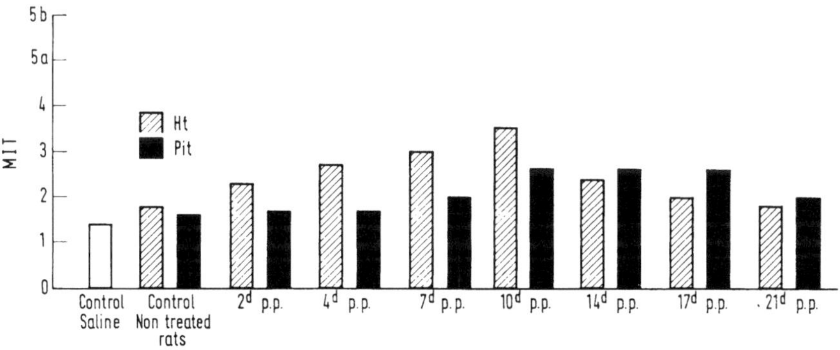

Fig. 45. Natural development of PIF suppression and PRF stimulation in nursing rat mothers: Average Mammotropic Index (MTI) of rats after 10 days' priming with estradiol (8 μg/d), followed by 5 days' injections of homogenates from hypothalami (Ht) or pituitaries (Pit) taken from nursing rat mothers 2—21 days post partum (p. p.). On the 10th day of nursing the hypothalamic PRF arrives at its peak, and is followed by a protracted peak of pituitary prolactin on days 10—17

A day-dependent curve of mammotropic index was obtained for the homogenates of hypothalami and pituitaries from post-partum rats injected into primed rats. The peak of these curves occurred on the 10th day post-partum in regard to the hypothalamic homogenates, and around the 14th day for the pituitary glands, i. e., maximal prolactin releasing activity appears in the hypothalamus on the 10th day post-partum, stimulating the pituitary gland to peak secretion of prolactin at the 10th—17th day (Fig. 45).

In conclusion it can be stated that these findings point towards impairment of the balance between PIF and a tentative PRF, indicating that the latter may be strongly stimulated during natural lactation. As to the part that oxytocin plays in this complex, it was believed for many years that oxytocin is the milk "let down" hormone by both central and peripheral mechanisms. From our results it seems that oxytocin, even in high doses (0.1—10 IU), does not release LTH from the pituitary, and therefore has no effect on mammary gland development. The final answer to this complex problem can only be established by localization of the hypothalamic site responsible for active prolactin release, using oxytocin retard preparations, able to supply a continuous release of small doses of oxytocin.

Chapter 24

Prolactin as Luteotropic Hormone

The ability of pituitary extracts to prolong the normal estrus cycle in rats and mice had been demonstrated (LONG and EVANS, 1922; TEEL, 1926) before the recognition of prolactin as a pituitary hormone (STRICKER and GRUETER, 1928, 1929). Later, prolactin was found to be responsible for prolongation of the duration of diestrus in the mouse (DRESEL, 1935) and rat (LAHR and RIDDLE, 1936). ASTWOOD (1941) found that, in hypophysectomized rats, injections of prolactin could produce functional corpora lutea. Prolactin induces pseudopregnancy in hypophysectomized rats and leads to the formation of deciduomata in a damaged uterine horn (EVANS, SIMPSON, LYONS and TURPEINEN, 1941). CLEMENS, SAR and MEITES (1968) have shown that the luteotropic effect of pituitary prolactin is necessary, during the first 6 days of pregnancy in the rat, to maintain progesterone secretion by the corpora lutea.

Homologous grafts of the anterior pituitary beneath the kidney capsule, which have been shown to secrete large amounts of prolactin, will also maintain the corporea lutea of rats in a functional state for long periods (EVERETT, 1954, 1956; NIKITOVITCH-WINER and EVERETT, 1958; DESCLIN, 1956). Reserpine has been shown to produce lactation in rats and at the same time to maintain the corpora lutea in a functional state (DE FEO, 1957; DESCLIN, 1958; BARRACLOUGH and SAWYER, 1959). The same effect has been shown for chlorpromazine (BARRACLOUGH and SAWYER, 1959) and for perphenazine (BEN-DAVID, 1968 b).

Luteotropic activity can, thus, be observed directly on the corpora lutea, either by gross examination (KOVACIC, 1968) or by histology, where the cells of the corpus luteum increase in size under the influence of prolactin (EVERETT, 1944; MANDL, 1959; WOLTHUIS, 1963). Another criterion could be the indirect result of lutein stimulation, as evidenced by prolongation of diestrus, induction of pseudopregnancy and deciduoma formation and the histological appearance of the uterus and vagina—all being the result of progesterone secretion. In fact, a luteotropic hormone has been defined as one capable of evoking progesterone secretion (ASTWOOD, 1953; GREEP, 1961). This problem was studied by BEN-DAVID and SULMAN (1966) who found that, while prolactin was effective in priming the endometrium by promoting secretion of progesterone from functional corpora lutea, it was quite unable to bring the progestational endometrium to the characteristic decidual stage (Fig. 46). For deciduoma formation an additional stimulus is needed, probably one associated with histamine release, e. g. pyrathiazine (SHELESNYAK, 1959).

A proper balance in secretion of estrogen and progesterone is necessary for maintenance of induced uterine decidual reactions and vaginal mucification in pseudopregnant rats. A study was, therefore, undertaken by JOSIMOVICH (1968) to determine whether primate luteotropins, human placental lactogen (HPL) and human chorionic gonadotropin (HCG) could maintain this balance in the absence of the pituitary gland. He found that combined therapy with HCG and HPL supported a normal decidual reaction in the traumatized uterine horn and vaginal mucification for 7 or 8 days after uterine trauma. It was concluded that HPL acted in conjunction with HCG treatment of hypophysectomized pseudopregnant rats to provide normal duration of induced uterine decidual reactions and vaginal mucification.

It appears that, in rats and mice, some action by LH on the corpora lutea may be essential before prolactin can stimulate luteal function. In the rat, prolactin does not *in vitro* influence glucose utilization (ARMSTRONG, KILPATRICK and GREEP, 1963) nor does it stimulate progesterone synthesis (ARMSTRONG, O'BRIEN and GREEP, 1964) by corpora lutea from pseudopregnant rats. LH, however, stimulates glucose metabolism and progesterone synthesis by these ovaries. GOSPODAROWICZ and LEGAULT-DEMARE (1963) found that neither prolactin nor HCG had an appreciable effect on incorporation of labeled acetate by rat corpora lutea *in vitro*, but that both hormones given together significantly increased acetate incorporation. These studies suggest that both prolactin and LH may be necessary for stimulation of luteal function in rats. It is of interest in this connection that prolactin is apparently necessary for ovarian depletion of ascorbic acid in response to LH (GUILLEMIN and SAKIZ, 1963).

The question arises for how long can prolactin exert a luteotropic action in rats. EVERETT (1954, 1956) and DESCLIN (1956) observed that in rats hypophysectomized during estrus, when new corpora

Response rating	Weight range (mg)	Deciduoma response
0	150— 300	
$\frac{1}{2}$	300— 400	
1	400— 500	
2	500— 700	
3	700—1000	
4	1000—1500	
5	1500 or more	

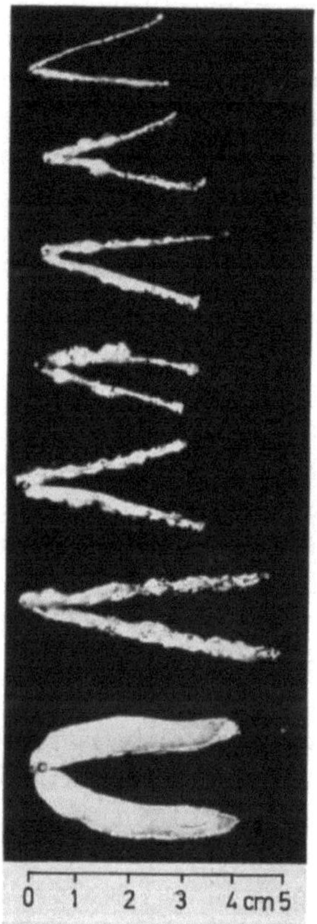

Fig. 46. Different degrees of deciduoma formation in the rat uterus after administration of prolactin (0.1—10 IU/rat), followed by pyrathiazine (1—20 mg/rat): 0, $^1/_2$, 1—negative reaction; 2, 3, 4, 5—positive reaction. The degree of the reaction was not proportionate to the doses of prolactin injected. It corresponded, however, to the doses of pyrathiazine given. This shows that the luteotropic effect of prolactin is negligible for the decidual responses

lutea are present, these are maintained in a functional state more or less indefinitely when a pituitary is grafted into the kidney capsule. On the other hand, MALVEN, HANSEL and SAWYER (1967) found that they could not prolong pseudopregnancy in rats by either prolactin or LH. MEITES and SHELESNYAK (1957) observed that pregnancy in rats could be prolonged by injection of prolactin. It is possible that the estrogen present during the latter phase of pseudopregnancy in the rat is insufficient for prolactin to prolong it.

In addition to their luteotropic action, prolactin injections have recently been reported to exert a luteolytic effect in rats (MALVEN and SAWYER, 1966). The authors noted that injections of prolactin, first given several weeks after hypophysectomy, hastened morphological regression of corpora lutea in adult rats. They postulated that the corpora lutea lose their capacity to be functionally maintained by prolactin within a few days after hypophysectomy. Earlier, ROTHCHILD (1965) had presented evidence that LH can exert a luteolytic effect in rats. Thus prolactin and LH may exert either luteotropic or luteolytic effects in rats, depending on the particular state of the corpora lutea.

In vitro studies (HUANG and PEARLMAN, 1962) failed to demonstrate any influence of prolactin on progesterone synthesis in the rat corpus luteum, whereas *in vivo* the luteotropic activity of this hormone is considered proven beyond doubt. Under similar experimental conditions, LH induced progesterone synthesis, thus fitting into the definition of luteotropic hormone.

In the ferret, pituitary stalk section resulted in the persistence of luteal function, suggesting that prolactin is luteotropic in the ferret as well as in the rat and mouse (DONOVAN, 1963 b).

In the ewe, in which formation of corpora lutea had been artificially induced, prolactin caused progestational activity (MOORE and NALBANDOV, 1955). Similarly, intact ewes, infused with prolactin into the ovarian arteries, showed enhanced progesterone secretion into the ovarian veins (DOMANSKI and DOBROWOLSKI, 1966).

As for the cow, controversial reports have been published on the luteotropic activity of prolactin. MASON, MARSH and SAVARD (1962) reported that LH, when added to bovine corpus luteum slices incubated *in vitro*, induced an enhanced synthesis of progesterone. Prolactin, however, under similar conditions failed to enhance progesterone synthesis, i. e., did not prove luteotropic. A similar observation was reported by ARMSTRONG and BLACK (1966) and by others. In a more recent work, however, prolactin as well as LH were shown to augment progesterone synthesis and secretion rate in bovine luteal ovaries perfused *in vitro* (BARTOSIK, ROMANOFF, WATSON and SCRICCO, 1967).

In the rabbit, prolactin failed to maintain the corpora lutea after hypophysectomy (MAYER, 1951), but LH did (KILPATRICK, ARMSTRONG and GREEP, 1964). On the other hand, HILLIARD and SAWYER (1966) reported that, after continuous infusion of LH into rabbits *in vivo,* the initial enhanced progesterone secretion rate was not maintained. On this background of continuous LH infusion prolactin did enhance the progesterone secretion rate. In a subsequent study the same investigators demonstrated that, while LH accelerated the synthesis of progesterone in the rabbit ovarian interstitial tissue, cholesterol stores were depleted. Thus, steroidogenesis was subsequently depressed. The addition of prolactin at this stage promoted cholesterol storage and restored the sensitivity of the ovary to LH (HILLIARD, SPIES, LUCAS and SAWYER, 1968). In addition prolactin maintained the morphological structure of the interstitial tissue, the significance and function of which remain to be clarified.

In the hamster, GREENWALD (1967) demonstrated that functional luteal maintenance in the pregnant animal involves a hormone complex rather than a single factor. Synergism of FSH and prolactin was found to be a prerequisite for luteal maintenance in this species. Thus, in hypophysectomized hamsters with ectopic pituitary grafts secreting prolactin, luteal function was not maintained without additional administration of FSH (CHOUDARY and GREENWALD, 1967). Conversely, the addition of large doses of LH along with prolactin and FSH terminated pregnancy in pregnant hypophysectomized hamsters (GREENWALD, 1967).

Recent observations may eventually induce us to change our concept of a single luteotropic hormone and think in terms of a "luteotropic-hormone-complex"—the result of synergism between two or more pituitary tropins, and possibly estrogen as well (SPIES, HILLIARD and SAWYER, 1968). Such a synergism would explain the failure to demonstrate luteal maintenance and progesterone secretion from the corpus luteum by prolactin in many species, and the discrepancies between different research works. Animals with intact pituitaries react differently from hypophysectomized animals, and these, in turn, unlike isolated ovaries, e. g. *in vitro.*

There may be other reasons for failure to demonstrate the luteotropic activity of prolactin. (i) Faulty technique: BARTOSIK et al. (1967) claim that plain incubation *in vitro* might be unsuitable for the induction of progesterone secretion from luteal tissue and that in all the studies, in which such an effect has been reported, the ovaries were perfused, either *in vivo* or *in vitro*. They further suggest that "the stimulatory effect of prolactin depends on an interaction between the luteal cells and surrounding ovarian tissue." This last statement is further supported by the discovery that prolactin may exert a "tropic" effect on the ovarian interstitial tissue and not directly on the corpora lutea (HILLIARD et al., 1968). (ii) The luteotropic effect of prolactin may be subject to species specificity. (iii) The life span of the corpus luteum has been shown to be limited (ASTWOOD, 1941), hence its responsiveness to prolactin would be limited to a short period after ovulation, and any experiment carried out at a different time is doomed to failure. (iv) There always remains the possibility that prolactin is in fact not luteotropic in one species or another. In the rat luteotropic effects can be obtained by local depletion of monoamines in the median eminence of the hypothalamus (VAN MAANEN and SMELIK, 1968).

To conclude, the luteotropic activity of prolactin has been clearly demonstrated in the rat and mouse only. For other species, some evidence of the luteotropic effect of prolactin has been brought forward in the ferret (DONOVAN, 1963 a, b), the ewe (MOORE and NALBANDOV, 1955), the cow (BARTOSIK et al., 1967) and the rabbit (HILLIARD et al., 1968). Other studies have, however, failed to show any effect of prolactin on the luteal phase of the menstrual cycle in either the monkey (HISAW, 1944) or women (KUPPERMAN, FRIED and HAIR, 1944).

Part III

Prolactin Assay

Chapter 25

In vitro Methods

a) Effect of Hypothalamus

Earlier experiments on *in vitro* culture of mammalian pituitary glands have shown that these gradually lose their ability to secrete tropins into the nutrient medium. This was first proved for ACTH and subsequently for TSH and the gonadotropins (Guillemin and Rosenberg, 1955). Hormonal elaboration *in vitro* could, however, be restored after 3—4 days by the addition of hypothalamic extracts (Guillemin, Hearn, Cheek and Housholder, 1957). These findings reconfirmed the stimulatory hypothalamic control of the anterior pituitary hormones. Most of the modern experiments were carried out in an artificial medium M-199 (Difco). The technique is not complicated (Figs. 47, 48).

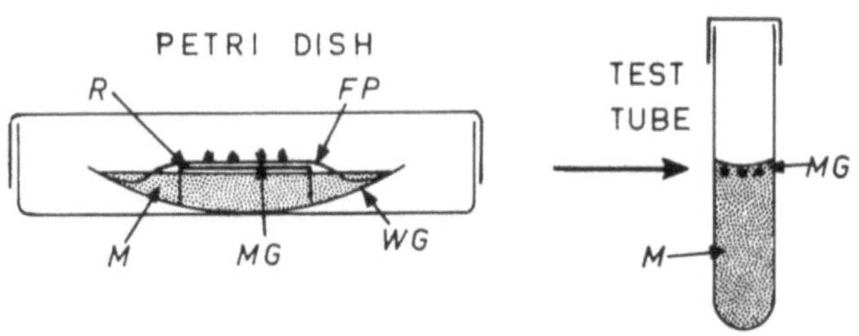

Fig. 47. Co-culture of hypothalamus, hypophysis and crop gland or mammary gland in medium 199: Petri dish; watch glass with plastic raft, filter paper, organ explants in medium 199. The Petri dish can be replaced by a test tube, since fatty organs float high enough for adequate oxygen supply.—*M* Medium (3 ml); *MG* Mammary Gland Fragments; *WG* Watch Glass; *R* Raft; *FP* Filter Paper

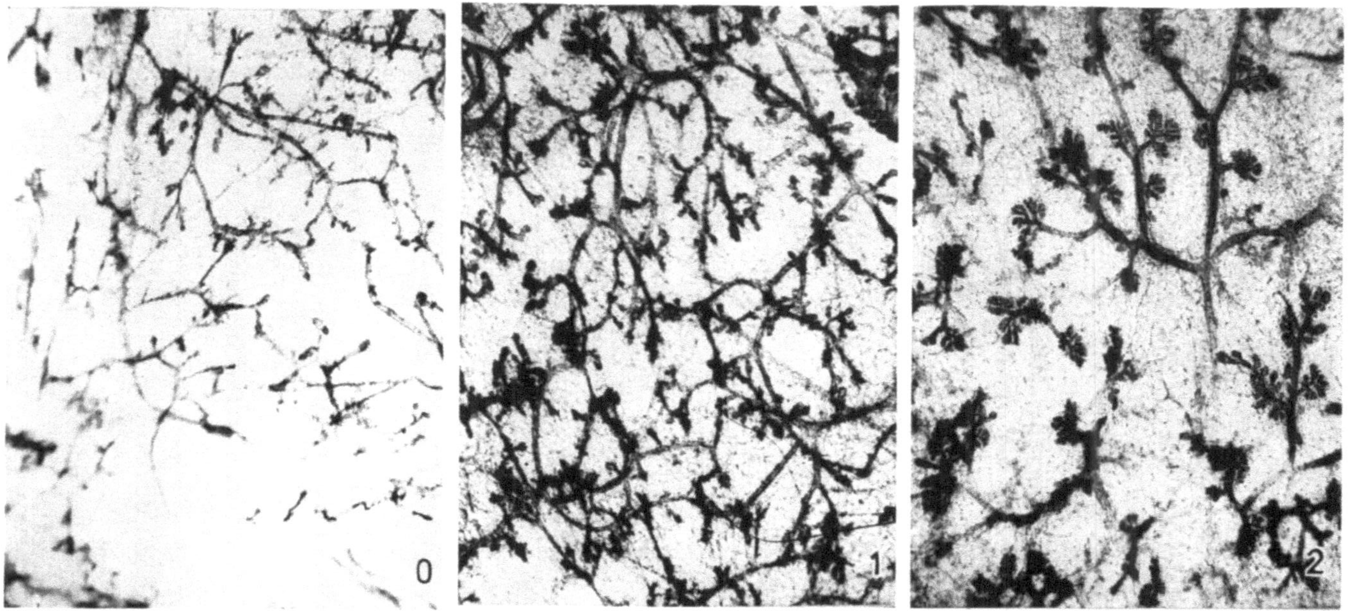

Fig. 48. MTI scoring *in vitro: In vitro* mammotropic index of mammary glands of female rats, weighing 270 ± 10 g, after fixation with Bouin's solution and staining with hematoxylin (24×).—MTI 0 represents normal gland pattern; MTI 1 shows stimulation by estrogen—"budding"; MTI 2 represents alveolar formation due to prolactin

b) Effect of Pituitary

Many *in vitro* experiments, such as pituitary transplantation, stalk section or the introduction of specific hypothalamic lesions, have shown that under normal conditions prolactin release is permanently inhibited by the hypothalamus. Consequently, the pituitary could be expected to secrete prolactin *in vitro* even in the absence of hypothalamic influence. This assumption was confirmed by the results of PASTEELS (1961 b) and of MEITES, KAHN and NICOLL (1961), who concurrently, though under different experimental conditions, cultured rat pituitary explants *in vitro*. The explants may preserve their structure well for 4 weeks. In these organ cultures the anterior pituitary secreted large amounts of prolactin for up to 33 days after explantation (PASTEELS, H. BRAUMAN and J. BRAUMAN, 1963). This secretory activity was increased, rather than decreased, in the course of cultivation, and was shown to amount to several times the quantity of prolactin present in the pituitary prior to explantation. Thus the autonomous ability of the anterior pituitary to secrete prolactin *in vitro* was established (ROBBOY and KAHN, 1966).

In contradistinction to other pituitary hormones, the addition of hypothalamic fragments to pituitary cultures *in vitro* significantly decreased the amount of prolactin secreted into the medium (PASTEELS, 1961 a; DANON, DIKSTEIN and SULMAN, 1963), while other nervous tissue was inactive. This finding suggested the presence of an inhibitory factor in the hypothalamus. A further step was the demonstration that this hypothalamic inhibitory effect was mediated by a humoral—non-nervous—factor, i. e. it is independent of the presence of intact hypothalamic tissue *in vitro*. Thus PASTEELS (1962) and TALWALKER, RATNER and MEITES (1963) showed that hypothalamic homogenates or acid extracts could elicit the same inhibitory effect on prolactin secretion as the hypothalamic tissue itself.

c) Effect of Steroids

Anterior pituitary cultures were also shown to respond to stimulation by hormones added *in vitro* to the culture medium. Thus, NICOLL and MEITES (1962 a, 1963, 1964) demonstrated that estradiol stimulates the anterior pituitary directly when added to an *in vitro* culture, resulting in an increased rate of prolactin release. Other hormones, however, did not display any direct effect on the *in vitro* cultured pituitary.

Both GALA and REECE (1964 b), using organ culture techniques, and PASTEELS (1963), using tissue culture methods, were unable to demonstrate any effect of estrogen on prolactin secretion by rat adenohypophysis *in vitro*. The reasons for the discrepancy between the negative *in vitro* findings of these investigators and the positive ones *in vivo* (KANEMATSU and SAWYER, 1963 b; RAMIREZ and McCANN, 1964) and *in vitro* (RATNER, TALWALKER and MEITES, 1962; NICOLL and MEITES, 1962 a, b, c; BEN-DAVID, DIKSTEIN and SULMAN, 1964; NICOLL and MEITES, 1964; RATNER and MEITES, 1964) are not clear, but may be found in different procedures employed.

BEN-DAVID, DIKSTEIN and SULMAN (1964) explored the effects of other steroid hormones on the hypothalamus-pituitary axis *in vitro*, using the technique described above. They found cortisone and cortisol to be strong stimulants of prolactin release in organ co-culture, apparently acting directly on the pituitary. High doses of cortisol, applied by NICOLL and MEITES (1962 a, 1963, 1964), depressed prolactin. Progesterone inhibited prolactin release in our set-up, again acting directly on the pituitary, while testosterone at the dose level used did not affect prolactin release. SAR and MEITES (1968) found that progesterone, testosterone propionate and cortisol acetate decreased hypothalamic PIF content, increased pituitary prolactin concentration and induced mammary growth. Testosterone propionate and cortisol acetate also elicited mammary secretion.

Norethinodrel (Enovid) also reduces the PIF, thus yielding increased prolactin release. This effect is apparently due to its being metabolized into an estrogenic compound (MINAGUCHI and MEITES, 1967).

These results indicate that, of all steroids employed, only estradiol and cortisol were capable of directly stimulating the pituitary to increase prolactin release. This is in complete agreement with the many *in vivo* studies in which administration of moderate but not large doses of estrogen or cortisol increased pituitary prolactin content and initiated milk secretion. Since estradiol and cortisol can also deplete the hypothalamus of PIF, the probability exists that under *in vivo* conditions they act on both the pituitary and the hypothalamus to promote prolactin release. Our observation of a direct stimulation by estrogen of prolactin release from the rat anterior pituitary *in vitro* is in agreement with the results of Nicoll and Meites (1962 a, 1963, 1964) but not with those of Gala and Reece (1964 b). Estradiol has been found to increase also prolactin release by the hamster pituitary *in vitro* (Nicoll, 1965). When incubated *in vitro*, the pituitaries of rats receiving estradiol injections release greater amounts of prolactin than pituitaries of control rats not injected with estradiol (Ratner and Meites, 1964). Pituitary prolactin content in cycling rats (Reece and Turner, 1937) and guinea pigs (Reece, 1939) is highest during proestrus and estrus and lowest during metestrus and diestrus (Meites, 1966).

d) Effect of Neurohormones

Trying to assess the effects of drugs and hormones on the hypothalamus-pituitary axis, Meites (1966) and his collaborators published a series of papers in which the drug under investigation was injected into the donor rats *in vivo* and the degree of inhibitory or stimulatory effect of the hypothalamic extract from treated animals on the pituitary *in vitro* was measured. They further introduced a short-term incubation technique, in which pituitary fragments were incubated and shaken for 2—4 hours in a Dubnoff shaker at 37° C under continuous O_2—CO_2 mixture gasing. By these techniques they found that adrenaline and acetylcholine (Mittler and Meites, 1967) reduced the hypothalamic prolactin-inhibitory factor, thus yielding a higher prolactin output from anterior pituitary cultures subjected to hypothalamic extracts from pre-treated animals. Gala and Reece (1965 a, b) found that adrenaline (and crude oxytocin) increased prolactin release. These findings require re-examination.

e) Effect of Phenothiazines

A different approach was taken by Danon, Dikstein and Sulman (1963). They tested the effect of the phenothiazine derivative, perphenazine, on a combined hypothalamus-pituitary organ culture and showed that perphenazine, added *in vitro*, could reduce or offset the effect of the hypothalamic prolactin-inhibiting factor. Perphenazine did not have any significant effect on the anterior pituitary itself, but the perphenazine-treated hypothalamus did not inhibit prolactin release. Thus, a depressant effect of perphenazine on the hypothalamic prolactin-inhibiting factor (PIF) was demonstrated *in vitro*. For other experiments, in which similar pharmacological effects of drugs on nervous tissue were studied *in vitro*, the reader is referred to a review by M. R. MURRAY (1965).

f) Effect of Reserpine

RATNER, TALWALKER and MEITES (1965) compared the effect of reserpine on the hypothalamus after its injection into rats and its addition *in vitro* to explants of the hypothalamus and pituitary incubated for 2 hours in a Dubnoff shaker. In both cases they found the same degree of depletion of hypothalamic PIF. This clearly indicates that the effect of the drug on the hypothalamus is similar *in vivo* and *in vitro*. NICOLL (1970) has, however, shown that MEITES' method of short incubation may yield false results. During the first 4 hours the hypothalamus produces a PIF, but later it produces a PRF which may give opposite results.

g) Effect of Thyroid

The profound effect of thyroid hormones on adenohypophysial function has long been recognized by endocrinologists. Although injections of thyroid hormones in rats can increase prolactin secretion by the adenohypophysis *in vivo* (MEITES, NICOLL and TALWALKER, 1963), the locus of action of the thyroid hormones was not established by these *in vivo* studies. NICOLL and MEITES (1963) observed that addition of either thyroxine or triiodothyronine (both at 0.1 μg per ml) to the medium of organ cultures of rat adenohypophysis resulted in a pronounced stimulation of prolactin secretion. A direct action of the thyroid hormones on the adenohypophysial prolactin cells was thus demonstrated. Evidently, of all the hormones tested, only estrogens and thyroid hormones were capable of stimulating prolactin secretion in the rat by a direct action on the adenohypophysial prolactin cells. In this context, it is pertinent to mention that COHERE, BOUSQUET and MEUNIER (1964) found that both thyroxine and estrogen stimulated development of the endoplasmic reticulum of the prolactin-producing cells of explants of rat adenohypophysis in organ culture.

h) Effect of Oxytocin

The hypothesis of BENSON and FOLLEY (1956) that prolactin release is stimulated by oxytocin has aroused considerable interest and controversy since it was first advanced. *In vitro* evidence has since been offered which contradicts the view of BENSON and FOLLEY (1956, 1957). NICOLL and MEITES (1962 b) studied the action of oxytocin (Pitocin) and vasopressin (Pitressin) on prolactin secretion by the adenohypophysis of rats in organ culture. No effect was observed with either neurohypophysial preparation in three days of culturing, a finding confirmed by PASTEELS (1963). TALWALKER, RATNER and MEITES (1963) also found that neither oxytocin nor vasopressin nor any of several other neuropharmacological agents were able to influence prolactin release. Our results agree with these findings.

The content of the prolactin-inhibiting factor in the hypothalamus has been reported to vary in oxytocin-mediated situations which may alter prolactin release. It drops during lactation (RATNER and MEITES, 1964), although it is not altered detectably following acute suckling (GROSVENOR, 1965 c).

i) Conclusion

The hypothalamus—pituitary organ co-culture is an excellent tool for prolactin assay, if its effect is evaluated either in the crop gland test or by addition of mammary tissue to the culture medium. Addition of hypothalamus of normal animals impairs prolactin secretion from the pituitary into the medium. Hypothalamus from nursing animals, however, increases prolactin production. Prolactin output is increased by addition of estrogens, norethinodrel, hydrocortisone and derivatives of phenothiazine or reserpine; thyroxine and triiodothyronine have a similar effect. Prolactin secretion is suppressed by the addition of progesterone and of most of its derivatives. The effect of adrenaline and acetylcholine is not clear. Testosterone, oxytocin and vasopressin are not active on prolactin release in this set-up. Unfortunately, the quantitative application of this test encounters a gamut of difficulties represented by the factors mentioned in this Chapter. It begins with the problem of creating a sensitive target organ by priming with estrogen or other steroids followed by the interaction of thyroxine, insulin and oxytocin. Thus, the combined organ co-culture could well serve for the determination of the prolactin-releasing or inhibiting activity of drugs if a standardization of all factors required could be achieved.

Chapter 26

In vivo Methods

This synopsis contains only the few methods adopted by us.

a) Rabbit Tests

α) *Rabbit Intraductal Method* (LYONS' Nipple Test, 1942, Fig. 49)

Priming: (a) 100 μg estradiol injected s.c. daily for 10 days into female rabbits (3.0 ± 0.2 kg B.W.), or: (b) 100 IU of HCG in one single i.v. injection in female rabbits (2.5 ± 0.2 kg B.W.).

Injections: Prolactin test starts on 11th day: one single injection of 0.25 ml mixed with 0.25 ml of gelatin (5% solution 40° C), injected through a 26-gauge hypodermic needle into one duct of a nipple. Controls of 0.25 or 1 IU standard prolactin prepared similarly are injected into another duct of the same nipple.

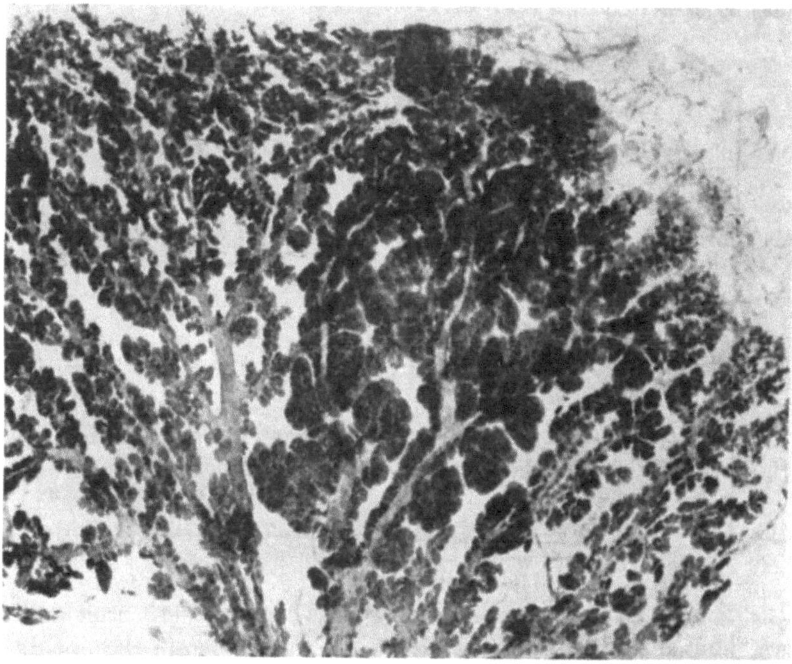

Fig. 49. Local prolactin assay in ovariectomized-hysterectomized virgin rabbits primed with estrogen and progesterone (by courtesy of W. R. LYONS, 1942, University of California, San Francisco Medical Center)

Assessment: After 3—4 days, comparison of the regions which contain milk, of both unknown and standard prolactin.

Evaluation: Inexact but convenient method, since 6—8 determinations can be performed on the same rabbit.

Sensitivity: 0.25—1 IU.

β) Rabbit Intracerebral Test (Stereotaxic Method) (MISHKINSKY, LAJTOS and SULMAN, 1966)

Priming: 100 μg estradiol injected s.c. daily for 10 days into female rabbits (3.0 ± 0.2 kg B.W.).

Technique: Implantation of microcrystals of prolactin-releasing substances in different sites of brain, especially hypothalamus.

Assessment: Copious lactation after 5 days.

Evaluation: Not quantitative, but convenient for localization of prolactin-inhibiting or releasing areas and for testing of lactogenic activity of unknown substance (Fig. 50).

b) Rat Tests

α) Mammotropic Index in Female Rats (TURNER, 1939)

Priming: Female rats (200 ± 10 g B.W.) receive not more than 8 μg/d estradiol s.c. daily for 10 days.

Injections: Prolactin test starts on 11th day: one s.c. injection of 0.2 ml daily for 5 days.

Autopsy: 16th day—rats killed, mammary glands fixed, mounted and stained (Fig. 51).

Evaluation: Test is highly specific if rats are correctly primed. Mammotropic Index (MTI) cf. Fig. 52. 1—2: neg., 3: ±, 4—5 pos.

Sensitivity: 2 IU exogenous prolactin, but very sensitive to endogenous prolactin release induced by drugs.

β) Milk Yield Test in post-partum Rats (GROSVENOR and TURNER, 1959 a)

Animals: Rat mothers (250 ± 20 g B.W.) with 6—10 litter mates each, housed in individual cages. No treatment before the 6th day post-partum.

Injections: Lactogenic preparations injected s.c. once daily during the 7th-20th day of suckling, saline controls required.

Suckling: Ten hours before suckling, litters are isolated from mothers which get 1 IU synthetic oxytocin s.c. immediately before suckling.

Milk yield: Estimated from pups' weight gain and mothers' weight loss during 1 hour or test suckling on days 14, 16, 18 and 20 postpartum.

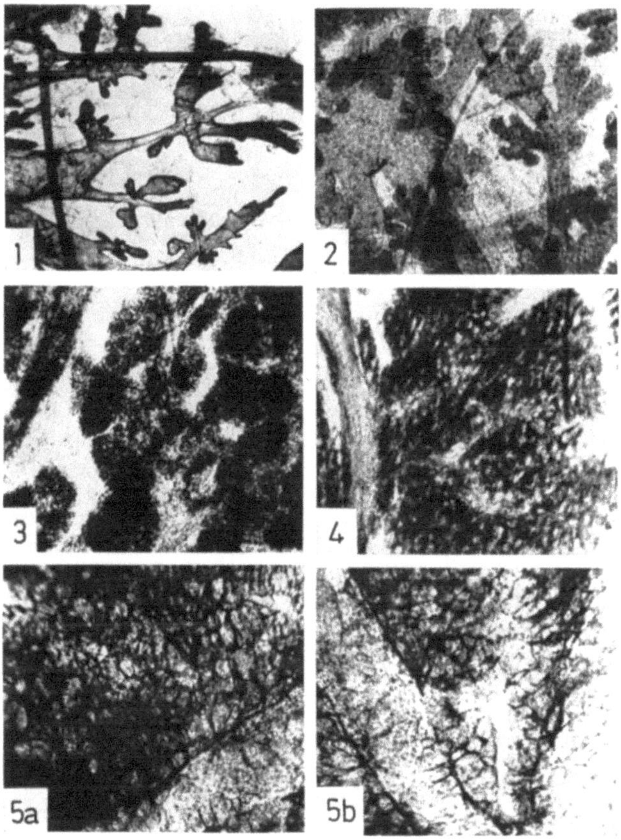

Fig. 50. MTI scoring in rabbit milk glands *in vivo:* Mammotropic index (MTI) of mammary glands of adult female rabbits shows different stages of development when examined in toto after fixation with Bouin's solution and staining with hematoxylin (12×). — MTI 1—2 represent the effect of ineffective drugs: normal gland pattern; MTI 3 represents gland after priming with chorionic gonadotropin; MTI 4—5 a—5 b represent increasing degrees of lactogenic effect (in 5 b the milk has been ejected)

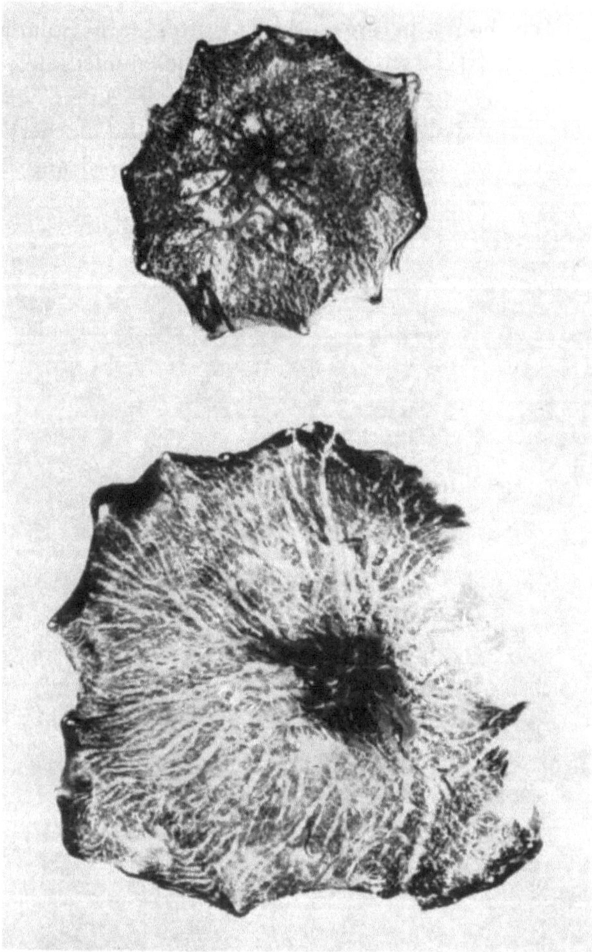

Fig. 51. Whole mount of rat mammary gland: small mount: normally developed gland of adult rat; large mount: same gland after induction of hypothalamic lactation by perphenazine

Evaluation: Test is not specific to exogenous prolactin because of surplus of endogenous prolactin. It allows a superior evaluation of the effect of inhibitory or auxiliary lactogenic factors.

Sensitivity: 20 IU prolactin.

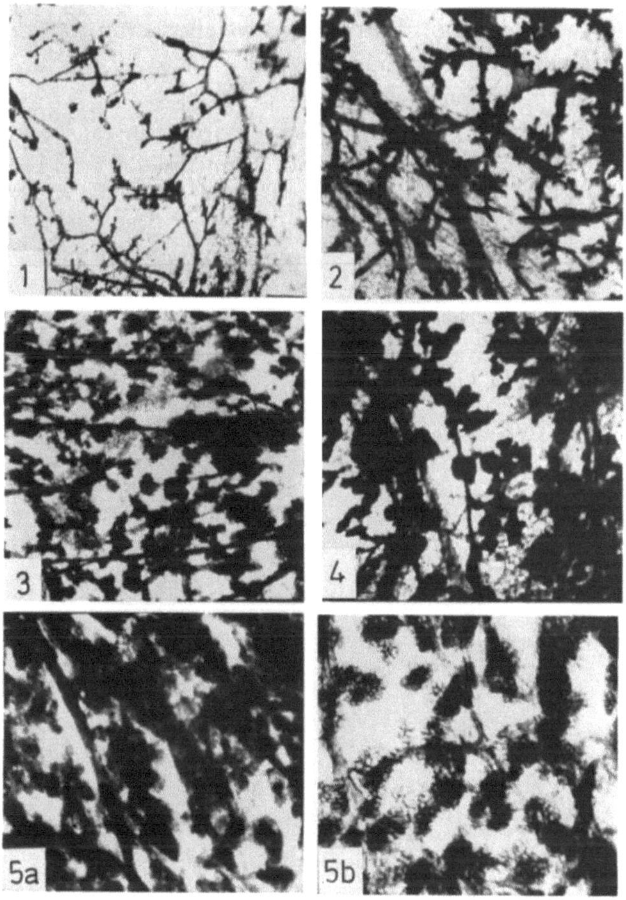

Fig. 52. MTI scoring in rat milk glands *in vivo:* Mammotropic index (MTI) of mammary glands of adult female rats in different stages of development, examined in toto after fixation with Bouin's solution and staining with hematoxylin (12×). MTI 1—2 represent the effect of estrogen and progesterone: normal gland pattern. MTI 3—5 represent increasing degrees of lactogenic effect. In 5 a the alveoles contain milk. In 5 b the milk has been ejected

c) Pigeon Tests

α) Systemic Pigeon Crop Gland Weight Method (RIDDLE, BATES and DYKSHORN, 1933)

Injections: 1 ml of each specimen injected daily s.c. into 4 pigeons for 4 days.

Autopsy: 5th day—birds killed and weighed, crop gland removed, spread, cleaned, blotted dry and weighed.

Evaluation: No practical use.

Sensitivity: 5 IU. Only high doses of prolactin can be estimated. High variability in response, depending on time, age and weight of the birds.

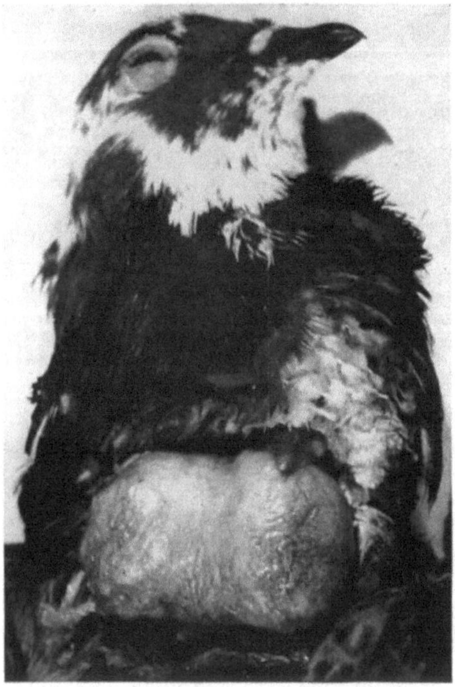

Fig. 53. Local crop gland response after intradermal injection: r: positive reaction; l: control reaction

β) Intradermal Pigeon Crop Gland Method (REECE and TURNER, 1937)

Injections: Minimal effective dose (0.05 IU prolactin in 0.05 ml normal saline) injected as standard in four equal parts during two days intracutaneously above one side of the crop gland. On the other side, dilutions of the material to be tested are injected in the same fashion (0.05 ml). The unknown material is diluted until the response is equal to that of the standard (Fig. 53). Addition of hydrocortisol (V. BEERSWORDT-WALLRABE, HERLYN and JANTZEN, 1965) or prednisolone (Herlyn, Jantzen, Flaskamp, Hoffmann and V. Beerswordt-Wallrabe, 1969) reduces unspecific inflammatory reactions. Blood prolactin can be determined by the method of Ben-David and Chrambach (1970).

Autopsy: 5th day. Assessment based on: ridging, wrinkling and measuring of the proliferative area (Fig. 54). The dry mucosa weight method is most reliable (Nicoll, 1967, 1969).

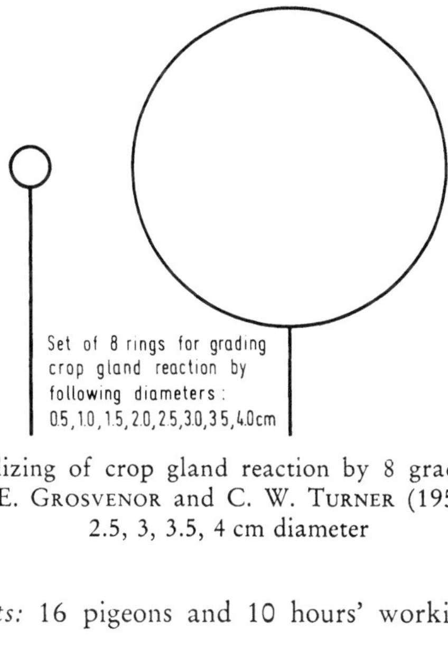

Fig. 54. Standardizing of crop gland reaction by 8 graduated wire rings, according to C. E. Grosvenor and C. W. Turner (1958): 0.5, 1, 1.5, 2, 2.5, 3, 3.5, 4 cm diameter

Requirements: 16 pigeons and 10 hours' working time for one titration.

Sensitivity: 0.05 IU, using 4 pigeons for each assay, differences of 50% are detectable. The H^3-methyl-thymidine uptake developed by Ben-David is more sensitive but expensive (Fig. 55).

d) Concentration Methods

A major problem in the local assay methods is the occurrence of non-specific inflammatory or other reactions, resulting from the injection of tissue extracts or body fluids. These reactions may be massive in the serosa and other submucosal layers and thicken the crop sac, thus imitating a prolactin response. The experienced investigator can, in most cases, distinguish between a true prolactin response and a purely inflammatory reaction. However, the problem remains where there are minimal responses or overlapping prolactin and inflammatory reactions. This difficulty is minimized by the use of mucosal weight according to NICOLL (1967), but some mucosal involvement also occurs and may be misleading (NICOLL and BERN, 1964). This shortcoming has been overcome by the addition of hydrocortisone or prednisolone, as mentioned above. BAHN and BATES (1956) claimed

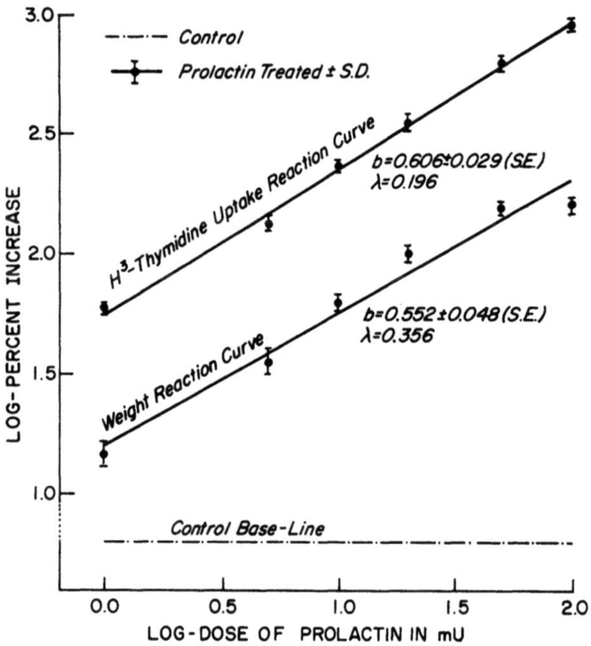

Fig. 55. Testing prolactin with tritiated thymidine. Relation between weight reaction curve of pigeon crop gland test and H^3-methyl-thymidine uptake

they could distinguish genuine prolactin reaction from non-specific responses by the cytoplasmic basophilia characteristic of the former and absent in inflammatory mucosal proliferation. Using this criterion, these authors could not support former reports (LYONS and PAGE, 1935; MEITES and TURNER, 1941; COPPEDGE and SEGALOFF, 1951) on the existence of measurable amounts of prolactin in blood and urine of human subjects.

SIMKIN and GOODART (1960) and SIMKIN and ARCE (1963) proposed using SULMAN's acid-acetone extraction method of chromatophorotropic activity (1952) for the extraction of prolactin from human blood, thus minimizing non-specific responses associated with the presence of foreign proteins. This procedure has been tested in our laboratory (unpublished data) and found unsatisfactory for the local crop-sac prolactin assay. A careful study of SIMKIN and GOODART's original paper reveals its non-specificity: in a recovery test, upon addition of 0.05 IU prolactin, they recovered ten times that activity as calculated on their own dose-response curve. The overlapping non-specific activity in this procedure is obvious.

Neither MEITES (1966) nor our group were able to confirm any of the results with rat or human blood claimed by SIMKIN and GOODART (1960). MOON (1966) also attempted to confirm their findings in rats, but without success. Inasmuch as the blood extracts employed elicit inflammatory reactions, it seems likely that the response measured by this technique is merely inflammation. Such non-specific inflammatory reactions may involve both the mucosal and submucosal tissues of the crop-sac (NICOLL and BERN, 1964), and can be elicited by diverse substances such as raw urine, mustard, and tannic acid (BAHN and BATES, 1956). We, therefore, tried to make use of the basophilia of the crop-sac cells in order to increase the specificity of the test by using basophil stains, as e. g. hematoxylin or methylenblue—unfortunately with negative results.

We presently use the following method for assaying prolactin in blood (BEN-DAVID and CHRAMBACH, 1970) (Fig. 56): The prolactin activity is extracted into the supernatant of serum brought to a concentration of 76% ethanol and 2% NaCl, according to CANFIELD and BATES (1965), at 0° C. Aliquots of serum are pipetted into 30 ml round bottom Corex centrifuges tubes. The appropriate amount of NaCl is dissolved in the serum by magnetic stirring. Cold absolute ethanol is added to a final concentration of 76%, the tube is covered with Parafilm, inverted several times, and the suspension stirred magnetically for 10 to 20 min. It is then centrifuged at 37,000 g for 20 min. Supernatants of aliquots of each pool or serum are combined in a 125 ml round bottom flask and subjected to flash evaporation, using a heating bath not exceeding 35° C and a "Rotovapor" connected to an oil pump with dry ice trap. Flash evaporation is terminated when the solution becomes cloudy. The cloudy solution is then transferred to

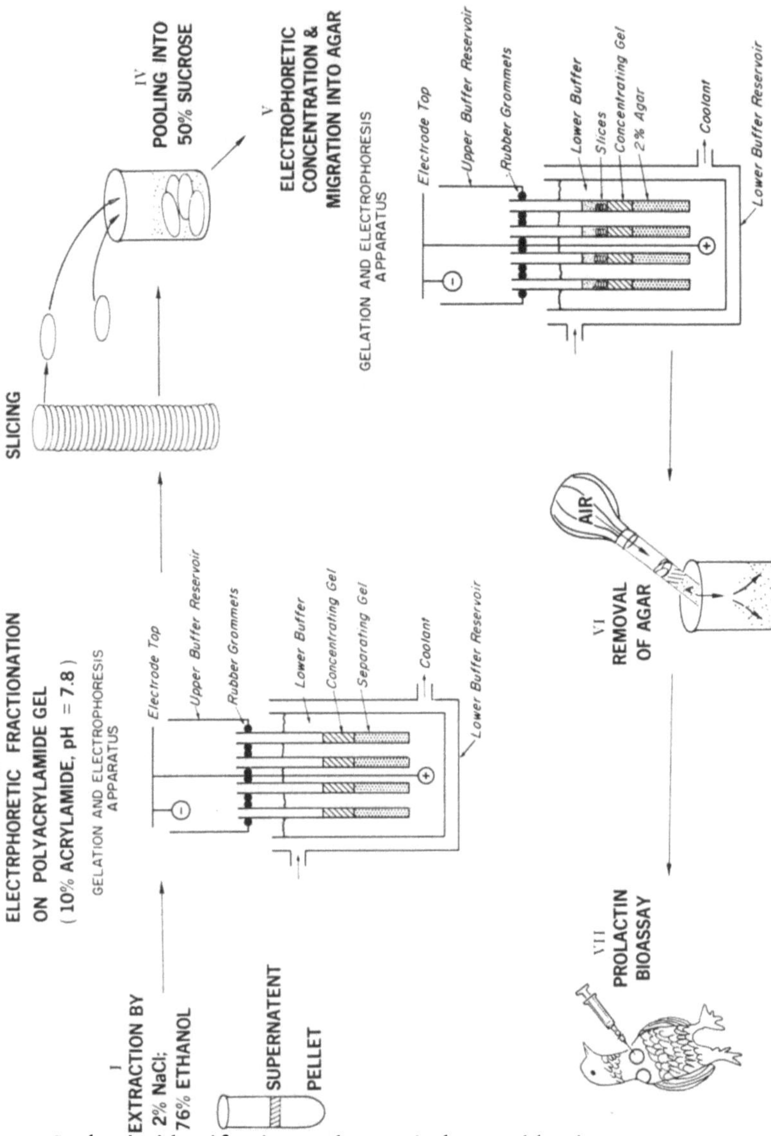

Fig. 56. Prolactin identification and assay in human blood: Stage I: extraction with 2% NaCl and 76% ethanol; Stage II: the lyophilized supernatant is subjected to electrophoretic fractionation on polyacrylamide gel; it can be injected directly into pigeons by the intradermal crop gland test. If prolactin identification and separation from STH is required—the following stages III—VII should be applied: Stage III: identification of active slices (Rf); Stage IV: pooling of active slices in 50% sucrose; Stage V: electrophoretic transport of prolactin into 0.2% agar; Stage VI: separation of agar from polyacrylamide gel by air blow; Stage VII: injection of agar prolactin concentrate into pigeons

dialysis tubing of 6.5 mm diameter[1]. The material is dialyzed at 4 °C against distilled water until the specific conductance of the dialyzate has reached the value obtained for distilled water. The dialyzed serum remnant within the tubing is lyophilized overnight and the product is stored at $-20°$ C. Prior to use the lyophilized serum extract is dissolved in 0.1 M Na_2HPO_4 with magnetic stirring. It is then neutralized by an equal volume of 0.1 M NaH_2PO_4. The total volume of phosphate buffer used is 0.2 ml per ml of serum. The equivalent of 1 ml serum is assayed in one pigeon and 5 pigeons are used for each assay point.

Chapter 27

Radioimmunoassay

Radioimmunoassays for rat, ovine and bovine prolactins were recently described (Fig. 57). KWA and VERHOFSTAD (1967) developed a radioimmunoassay for rat prolactin, using rat prolactin prepared from transplants of pituitary tumors as a reference antigen. These authors were able to detect changes in prolactin levels in blood of individual female rats through the estrus cycle and through pseudopregnancy by taking a 1 ml blood sample daily from the tail vein. High concentration of prolactin was found during the onset of lactation and unexpectedly low values during pregnancy (KWA, 1968). A similar method for rat prolactin has been described by NISWENDER, CHEN, MIDGLEY, MEITES and ELLIS (1969). Bovine and ovine prolactin do not appear to crossreact in this radioimmunoassay. Quantities of NIH-P-B2 and NIH-P-S8 as high as 1 μg/assay tube did not inhibit the reaction between the antirat prolactin and the rat prolactin-^{131}I. It appears that a radioimmunoassay specific for rat prolactin has been developed which is capable of measuring the prolactin in as little as 100 μl of serum from an immature rat and in 10 μl obtained from most females on the afternoon of proestrus. This radioimmunoassay

1 The dialysis tubing is washed as follows: The tubing is boiled in 10% Na_2CO_3 until no more yellow material appears in the wash; it is then washed to neutrality with distilled water and stored at 4° C in 0.01 M NaEDTA of pH 7.0. Just prior to use it is washed thoroughly with distilled water to free it from EDTA.

method is estimated to be about 1,000—2,000 times more sensitive than the local pigeon crop bioassay. It has been used to measure serum levels of prolactin in rats throughout their cycle, and when bearing prolactin secreting tumors. The purification of rat prolactin for these studies has been described by ELLIS, GRINDELAND, NUENK and CALLAHAN (1969).

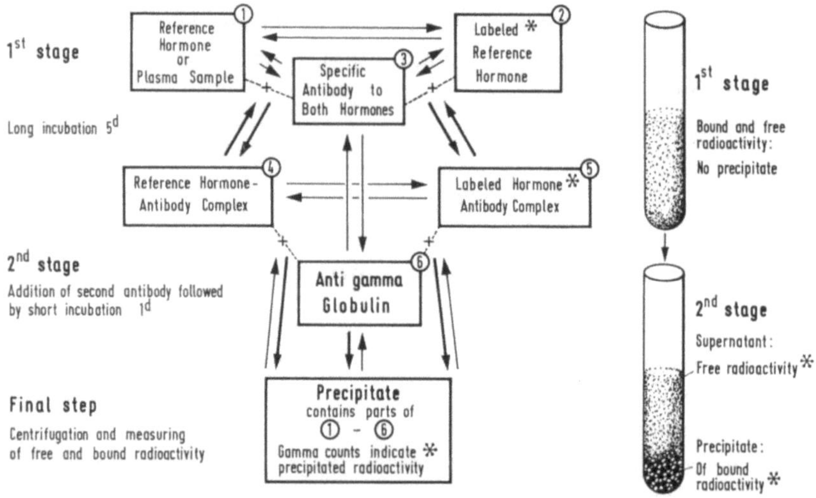

Fig. 57. Radioimmunoassay of prolactin. Diagram of "double-antibody" radioimmunoassay in three stages

ARAI and LEE (1967) reported use of a double-antibody radioimmunoassay for ovine pituitary prolactin. The ovine prolactin antiserum showed crossreaction with bovine prolactin but not with human preparation. Using the radioimmunoassay procedure, ARAI and LEE (1967) also found a gradual decline in prolactin levels during pregnancy from 100 mμg/ml in the normal non-pregnant sheep down to about 4 mμg/ml in the 4½ months of pregnancy. Subsequently they found a sharp rise post-partum and during lactation, up to 307 mμg/ml in the fourth month post-partum.

Plasma prolactin levels, detected in the goat by radioimmunoassay, were reported by BRYANT, CONNAN and GREENWOOD (1968) and BRYANT and GREENWOOD (1968). They found that delivery, milking and i.v. injections of oxytocin resulted in a pronounced increase in the plasma prolactin levels. BRYANT and GREENWOOD also measured the rate of disappearance of ovine prolactin from the plasma after its i.v. injection into a goat over a 2-hr period. The half-time of disappearance from arterial, jugular and mammary blood was found to be 19 min. Cows seem to be able to lactate without significant increase of prolactin levels (SCHAMS and KARG, 1969). The only peak discovered by them was 24 hours before parturition, when estrogen levels are high.

Radioimmunoassay is the most sensitive test. When radioimmunoassays are used for the determination of protein hormones, it is important to remember that they are based on immunological specificity rather than on biological activity. It is clear that the molecular structural requirements for biological and immunological activities are not identical. For example, it has been demonstrated that the N-terminal part of ACTH is needed for biological activity, whereas the C-terminal section is needed for immunological activity (IMURA, SPARKS, GRODSKY and FORSHAM, 1965). Moreover, we found that the three forms of ovine prolactin detected through radioimmunoassay show four bands at polyacrylamide gel electrophoresis. But only a single main band exerts the biological activity of prolactin (BEN-DAVID and CHRAMBACH, unpublished data). This could simulate cross reactions with other tropic hormones or hormones derived from other species or even unspecific cross-reactions and thus provide false positive results. BRYANT, TECH, GREENWOOD and LINZEL (1968) actually found cross-reactions between ovine, bovine and caprine prolactin which allows the ovine prolactin radioimmunoassay to be applied to all these three species.

The amply discussed question of whether or not "human prolactin" does exist as a molecule separate from human growth hormone (LARON, ASSA and APOSTOLAKIS, 1965) has not yet been resolved because, radioimmunologically, "human prolactin" is not separable from human growth hormone. Using the pigeon crop bioassay and polyacrylamide gel electrophoresis, we were able to demonstrate different bands for human growth hormone and most active human prolactin preparations, such as preparation HSI-087-E of WILHELMI or preparation 36-R8 of LARON, ASSA and APOSTOLAKIS (1965). We could not, however, differentiate between these bands when using human growth hormone radioimmunoassay since they cross-react with each other (CHRAMBACH and BEN-DAVID, 1970). It is essential, therefore, to run from time to time a bioassay alongside with the radioimmunoassay to provide a permanent check of the latter.

Part IV

Problems of Lactation

Chapter 28

Lactation in Humans

a) Lactation and post-partum Amenorrhea

For successful lactation it is essential that the patient be endowed with breast tissue that is capable of normal response to estrogens, progesterone, and lactogenic hormones (Fig. 58). The rising titer of the steroid hormones during pregnancy causes maximum development of the lobulo-alveolar system and presumably sensitizes the glandular tissue for the action of lactogenic hormones. Prolactin is secreted in increased amounts during pregnancy, but its release from the pituitary and the placenta is inhibited by the high estrogen titer. It should be pointed out that the colostrum, that may exude or be expressed from the breasts during pregnancy, contains little, if any, milk and consists chiefly of desquamated epithelium and transudate. It is not until about the third day post-partum, when the estrogens have been nearly completely cleared, that the breasts become engorged by the active secretion of milk. It is presumed that this is caused by the release of LTH and STH following the removal of the estrogen "brake." For lactation to continue actively, it is necessary that there should be regular suckling or artificial pumping of the breasts. This suggests the importance of the neural pathway from the breasts to the hypothalamus mediating the continued secretion of oxytocin and lactogenic hormones.

b) Inhibition of Lactation and Breast Engorgement

Hormonal therapy with estrogens or androgens has been found helpful in preventing painful engorgement of the breasts in those patients who do not wish to nurse, and in alleviating the engorge-

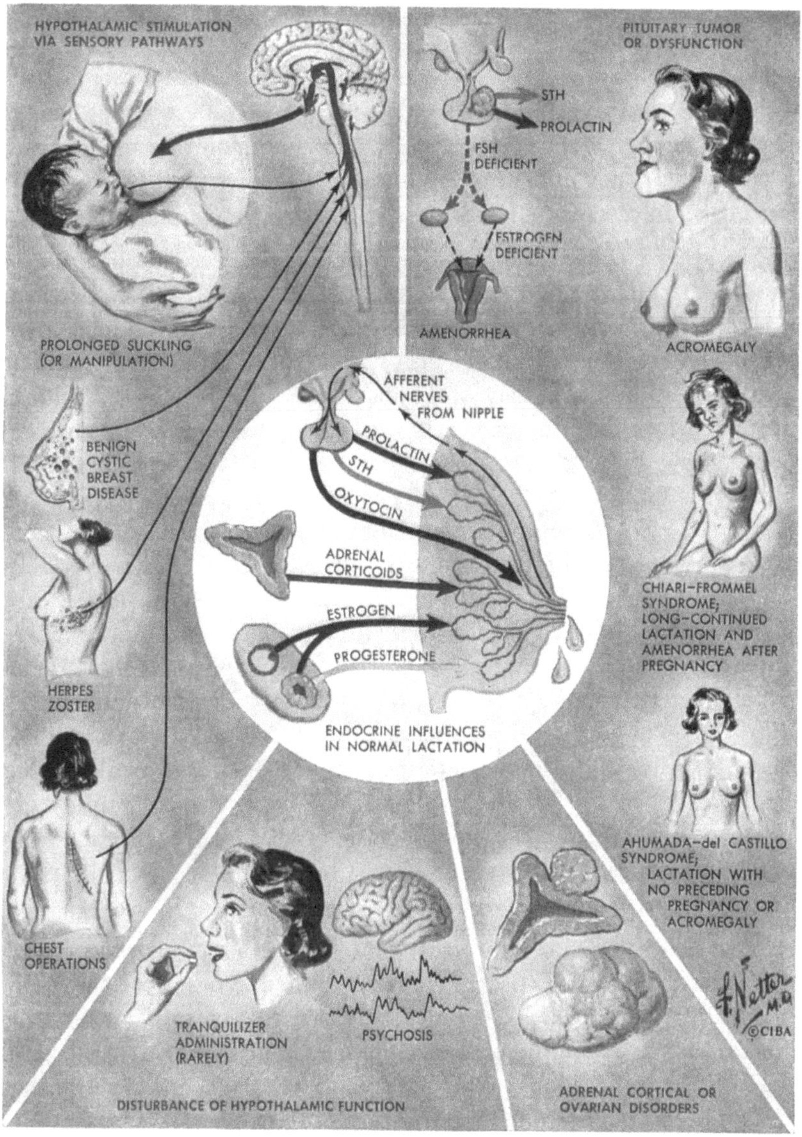

Fig. 58. Presentation of the various factors which may produce lactation or galactorrhea: Center: Afferent nerves from nipple, prolactin, STH, oxytocin, adrenal corticosteroids, estrogen, progesterone; Left side: Hypothalamic stimulation via sensory pathways operating possibly via the prolactin releasing factor (PRF), such as suckling, benign cystic breast disease, herpes zoster, chest operations. Lower part: Hypothalamic suppression by tranquilizers (including phenothiazines, reserpines, butyrophenones, etc.), in adrenal cortical or ovarian disorders, in Ahumada-del Castillo Syndrome and Chiari-Frommel Syndrome—all possibly operating via the prolactin-inhibiting factor (PIF). Right side: Pituitary tumor or dysfunction operating by direct release of STH and prolactin (acromegaly). (Copyright: The Ciba Collection of Medical Illustrations by Frank H. Netter, M. D.)

ment in patients who wish to wean their babies from breast feeding. Such therapy also helps in "drying up" the breasts if lactation has started, although some authorities are of the opinion that the decrease in milk flow is not influenced by hormonal therapy and is dependent entirely upon cessation of suckling.

Since 1937 it has been possible to inhibit the secretion of milk by the administration of different steroids. At first, injections of estradiol benzoate were used, but frequent ingestion of tablets containing small doses of estrone or estriol was also satisfactory. Lately, the synthetic estrogens—stilbestrol, dienestrol, and hexestrol, as well as ethinyl estradiol—have been more widely used owing to their low cost. Commonly, 5 mg stilbestrol six times a day for 4 to 10 days is given, but its effect is unpredictable according to HESSELTINE, BUSTAMANTE and NAVORI (1955). A combination of orally active, water-soluble, conjugated estrogens, obtained from the urine of pregnant mares (Premarin), and methyltestosterone is preferred by FISKIO (1955). The regimen is 3 tablets every 4 hours five times. While in about 40 per cent of the cases results are not good, there are no unpleasant side effects. It is claimed that similar combinations of ethinyl estradiol and methyltestosterone, having a synergistic suppressive action on lactation, also show less tendency to induce uterine bleeding on withdrawal. Tablets should be given at short intervals as these hormones are rapidly absorbed and excreted. The combination of 50 mg progesterone with long-acting estradiol benzoate (10 mg) by injection, given daily for 5 days, is more consistently effective.

Skepticism towards this treatment has been voiced recently by MACDONALD and O'DRISCOLL (1965). A double-blind trial of stilbestrol and placebo in 500 women was carried out to determine the effect of estrogens on discomfort from breast engorgement. Stilbestrol (10 mg/d) and placebo were almost equally effective, and it seems that the routine administration of small doses of estrogens to suppress lactation is not sufficient. Doses up to 100 mg/d may be required.

The use of a depot steroid mixture—Ablacton (Schering)—has recently been recommended for women. It contains estradiol benzoate (5 mg), estradiol valerianate (8 mg), ethinyl-testosterone acetate (20 mg) and testosterone enanthate (180 mg) and is given in one i.m. injection. This preparation is supposed to suppress hypothalamic PRF and pituitary LTH.

Alpha-ergocryptin suppresses lactation in rats, it has, however, not yet been tried in women (FLUECKIGER and WAGNER, 1968).

c) Persistence of Lactation and Menstrual Cycle

Persistence of lactation and spontaneous lactogenesis are rare conditions in modern civilized society. In western countries in the past, and even now in primitive races, it was economic policy for mothers and grandmothers to prolong lactation for many years by repeated suckling, so as to assist in feeding their large numbers of children and grandchildren. Some women secrete a little milk for a year or more after weaning, some do at the menopause; GESCHICKTER (1960) quotes a figure of 1 to 4 per cent of women in England and the United States with persistent lactation.

Spontaneous lactation (witch milk) often occurs in the breasts of newborn children and is no doubt due to the action of the maternal hormones. Many babies have slight breast development, and secretion of milk may develop after about a week. Their small breasts may be congested and hard. This usually subsides without trouble, but abscess formation sometimes appears.

In the normal lactating mother there is some degree of ovarian suppression, due probably to inhibition of pituitary FSH and LH. So, too, in cases of spontaneous galactorrhea, there is usually some associated estrogen deficiency manifesting itself in menstrual irregularity, hypomenorrhea, or incomplete development of the normal pattern of vaginal smear. There may, in addition, be varying degrees of atrophy of the genital tract and even of the ovaries themselves. Numerous reports have also confirmed a deficiency or absence of urinary FSH.

Evidence of overactivity of growth hormone exists in some of these cases, and acromegalic women sometimes have galactorrhea (Fig. 58). In 1954 FORBES et al. gave a full account of galactorrhea in 15 non-acromegalic women. The interesting finding was that in 8 of these patients there was evidence of a pituitary tumor, and in 3 biopsy confirmed the presence of a chromophobe adenoma. It is not known whether lactation is induced by overproduction of growth hormone and prolactin from the eosinophile cell tumor in the acromegalic or from the pre-eosinophilic cells of the chromophobe adenoma, or whether it is just a matter of pressure from a tumor or other stimulus affecting the hypothalamus. In all these cases, galactorrhea and amenorrhea were accompanied by absence of follicle-stimulating hormone excretion without any other signs of pituitary failure; in fact, the finding of hirsuties and elevated 17-ketosteroids in some cases would be consistent with hyperpituitarism.

d) Sterility during Lactation

The belief is current that lactation is associated with physiologic sterility, and also that it inhibits the return of uterine bleeding. The hormonal theory usually advanced is that these effects are due to pituitary prolactin suppressing the secretion of pituitary gonadotropins, especially LH.

Lactation is regarded generally to retard the return of uterine bleeding. De Lee (1924) observed that in women who do not nurse uterine bleeding usually returned within six weeks of delivery and recurred quite regularly. However, in nursing mothers, bleeding was said to return within three months in 50% of the women, in 71% within six months, and in 81% before nursing was discontinued. Mazer and Goldstein (1932) cite a study of 1,200 lactating women according to which 55% experienced a return of uterine bleeding while still nursing. These data indicate that the "amenorrhea of lactation" is but a relative affair and becomes uncertain after 3 months. During lactation, episodes of uterine bleeding occur at irregular intervals, and the flow is variable in amount.

The question whether ovulation occurs during "lactational amenorrhea" arises because of the occurrence of conception prior to resumption of the menses after delivery. A study by Griffith and McBridge (1939) of endometrial biopsy specimens, obtained from 21 lactating women during the 3rd to 24th week of puerperium, indicated that only one ovulated prior to the resumption of bleeding. In another study (Topkins, 1943) of 145 endometrial biopsy specimens, secured from 28 lactating women during the period of "lactational amenorrhea", only nine of these specimens (6%) showed progestational alterations, all of which were associated with the first episodes of flowing.

Other studies (Lass, Smelser and Kurzrok, 1938) indicate, however, that the early episodes of bleeding during lactation do not necessarily constitute true menstrual bleeding, i. e. they were not preceded by ovulation, corpus luteum formation or progestational proliferation of the endometrium. Forty-seven women, who were studied throughout their puerperia, had 194 regular episodes of bleeding, and 106 of these episodes (55%) were judged to have been preceded by sterile cycles since bleeding occurred from estrogenic endometria. Yet 82 episodes of bleeding (45%) occurred from progestational endometria and were thought to have been preceded by fertile cycles.

These different studies suggest 1) that ovulation does not generally occur during "lactational amenorrhea," and 2) that uterine bleeding during lactation may or may not be preceded by ovulation. Accordingly, sterility during lactation is relative and not absolute.

The advent of the contraceptive pills has made a more exact study imperative. CRONIN (1968) estimated that the non-fertile period in women who do not lactate lasts for 6 weeks after childbirth and in nursing mothers for 3 months. The validity of these estimates was checked by daily recordings of body-temperature in relation to menstruation in 93 non-lactators and in 81 women who did lactate. The mean time to first ovulation was 12 weeks in the 93 non-lactators, and a fertile ovulation preceded the first menstruation in 21 of them; there seems to be a 1-in-20 chance that a fertile ovulation will occur before the first menstruation and before the 7th week post-partum. Lactation extended the time lapse until the return of ovulation. Thus, provided breast-feeding is in progress and menstruation has not returned, ovulation before the end of week 10 post-partum will be a rare event.

Appreciation of the influence of lactation upon the resumption of ovulation after childbirth is important from the practical as well as the theoretical aspect. Definite knowledge of the earliest date at which ovulation can occur would make conception avoidance unnecessary before that date. The use of contraceptive methods—which may inhibit lactation—could thus be postponed to permit lactation to become well established—a consideration of great importance in countries in which the child depends on breast-milk for its intake of protein.

e) Galactorrhea as a General Syndrome

Galactorrhea (abnormal lactation) refers to mammary secretion of a milky fluid, which is nonphysiological in that it may be persistent, excessive, and unrelated to a recent pregnancy. The onset may date back to a normal post-partum lactation which did not cease (Fig. 58).

Amenorrhea which may, or may not, precede the lactation, genital atrophy, and other evidences of estrogen deficiency are frequent but not invariable phenomena. Pathological lactation occurs very seldom in females and is extremely rare in males.

The factors operative in abnormal lactation are not yet clear, especially because of the large variety of clinical conditions with which it is associated. In most instances, galactorrhea can be traced to an organic functional intracranial disorder, involving particularly the pituitary-hypothalamic structures. The final and effective link in the various pathogenetic mechanisms appears to be an unphysiological production of one or more adenohypophysial tropic hormones, i. e. LTH or STH. By further analogy with normal lactation, mechanical and psychic as well as humoral components seem to play a role. On the other hand, abnormal lactation is relatively rare in the various diseases occasionally associated with lactation.

Among the better known and more frequent causes of galactorrhea are pituitary tumors, with or without acromegaly. Eosinophilic, chromophobe, or basophilic neoplasms may be present. Galactorrhea and amenorrhea may be the first symptoms, so that, even when no tumor can be demonstrated, it is advisable to suspect such lesions in an incipient form. Other organic brain lesions include encephalitis, meningitis, and trauma.

Dysfunctional hypothalamus-pituitary disorders include the Chiari-Frommel syndrome which is characterized by galactorrhea, following pregnancy and nursing, in some cases due to unrecognized pituitary tumors, in others to primary ovarian failure; the Ahumada-del Castillo syndrome which is characterized by spontaneous galactorrhea, not related to pregnancy or acromegaly, sometimes due to pituitary neoplasm or to selective hyperpituitarism. The latter has recently been studied by SPELLACY, CARLSON and SCHADE (1968).

f) Forbes-Albright-Castillo Syndrome

This clinical entity which originates in the pituitary and is associated with lactation, has been described by FORBES, HENNEMAN, GRISWOLD and ALBRIGHT (1951, 1954) and by ARGONZ and DEL CASTILLO (1953). The clinical picture is characterized by the presence of amenorrhea (usually primary), a low FSH, and an acromegalic-like facies. Patients with this complaint usually have an enlarged sella turcica but do not show evidence of visual disturbances. They appear to be akin to frank acromegalics (who may also lactate), except that they do not show the organomegaly, increase in soft tissue of the hands, or marked encroachment on the visual fields usually associated with progressive enlargement of the sella. If menses are not desired in the patients with Forbes-Albright-Castillo syndrome, control of lactation may be accomplished sometimes by the use of 10.0 mg, three times daily, for 30 days, of norethindrone, norethandrolone, or norethynodrel. If lactation does not subside, the course of any of the 19-norsteroids should be continued beyond that period. The same schedule, using chlorotrianisene or clomiphene, may also be used with an equal degree of effectiveness where catamenia is desirable. Once lactation subsides, it does not usually reappear, but should it do so, another course of therapy may be instituted. Occasionally there are patients in whom lactation cannot be completely suppressed by this type of treatment and there is little that can be accomplished for these women since even twice the recommended dose of the 19-norsteroids would probably not be more effective either. In addition, when one increases the doses of the steroids to twice the initially recommended regimen, an increase in undesirable side effects may be noted. It should, however, be mentioned that the prospects for fertility of the lactating patients have become fairly hopeful since the introduction of clomiphene, which, however, does not unfortunately completely abolish lactation.

g) Chiari-Frommel Syndrome

In rare instances, amenorrhea and lactation may persist for many years in women who have had an otherwise normal post-partum period and have not developed any complicating organic disease. This syndrome was first described by Chiari, Braun and Spaeth in 1852, and later elaborated by Frommel (1852, 1855) (Fig. 58). It occurs most frequently in primiparas, who had, as a rule, uncomplicated deliveries. The patients include both nursing and non-nursing mothers. Menstruation fails to reappear within the usual period of time and, subsequently, amenorrhea persists indefinitely. The breasts continue to lactate, and in many instances a copious, milky discharge lasts for many years in spite of efforts to dry up the breasts. In other instances, the milky discharge is scant and demonstrable only by manual expression from the breasts.

On examination the breasts are found to be well developed and to contain abundant glandular tissue, but in a few instances milky secretion is present in patients with small flaccid breasts. The other striking finding is a marked degree of hypoplasia of the uterus. The endometrium is usually atrophic. The ovaries are small. Galactorrhea usually precedes amenorrhea, though the reverse condition may occur as well. It may persist for years together with amenorrhea, or there may be temporary intermissions in which presumably follicle-stimulating hormone is resecreted so that menstrual rhythm returns and pregnancy may ensue. The uterus may be of normal size or may show varying degrees of atrophy.

Endocrine studies have consistently shown absence of urinary gonadotropins (FSH), markedly diminished excretion of estrogens, and normal urinary 17-ketosteroids. In some cases there was diminished basal metabolic rate. Many of the patients complain of marked tiredness and increased fatigue. As a rule, no gross nervous or psychic disturbance seems to be involved.

Liggins and Ibbertson (1966) reported a successful quintuplet pregnancy following treatment with human pituitary gonadotropin in a patient with post-partum Chiari-Frommel syndrome. Preliminary investigation of the mother indicated that failure of ovulation resulted from a defect in pituitary-hypothalamic function. The response to clomiphene, which is assumed to cause a discharge of gonadotropin (Greenblatt and Mahesh, 1964—1965), indicated normal gonadotropin production. Although gonadotropin release was not measured, a failure of release was inferred from the monophasic temperature chart and the low urinary estrogen levels. Excessive prolactin secretion resulting from loss of the normal hypothalamic control (PIF) was obviously the reason for the persistent lactation in this case. The human-growth-hormone immunoassay technique used does not distinguish between growth hormone and prolactin, which are immunologically very similar (Berson and Yalow, 1964). Although STH levels were normal, the failure to suppress growth hormone to zero during a glucose infusion pointed to the possibility of an autonomous prolactin production.

The Chiari-Frommel syndrome is characterized by a marked resistance to therapy. It is usually difficult to induce bleeding even with large doses of estrogen and progesterone, but, once the first bleeding episode is induced, others can often be brought on more easily by cyclic administration of estrogen and progesterone. Subsequent resumption of spontaneous ovarian function is rare. Most patients also fail to react to gonadotropic stimulation. The addition of thyroid extract appears to be helpful in combating the fatigue but dose not influence the ovarian hypofunction. Clomiphene therapy is at present the best treatment, it reduces the milk flow, without, however, abolishing it entirely.

h) Sheehan's Syndrome

Selective hypopituitarism is a rare cause of galactorrhea, as is juvenile hypothyroidism with precocious menstruation. The possibility of compensatory pituitary adenoma cannot be excluded in these cases (MAAS, 1967).

i) Hypothalamic Lactation

Galactorrhea also rarely occurs in psychoses but often in patients receiving large doses of tranquilizers such as phenothiazine derivatives (cf. Chapters 14 and 15), rauwolfia alkaloids (cf. Chapter 16), butyrophenones (cf. Chapter 17) or methyldopa (cf. Chapter 18) (PETTINGER, HORWITZ and SJOERDSMA, 1963). As even imipramine (Tofranil) can evoke galactorrhea (KLEIN, SEGAL and PICHEL-WARNER, 1964), hypothalamic involvement is obvious. The pathogenetic mechanisms in prolonged suckling, benign cystic breast disease, dorsal herpes zoster or chest operations (SALKIN and DAVIS, 1949; GRIMM, 1955; QUINLAN, 1968) presumably involve stimulation of the sensory reflex arc from the chest region to the hypothalamus. The latter condition is especially resistant to steroid treatment. Other associated conditions include adrenocortical hyperfunction (tumor, hyperplasia), menopause, castration and ovarian tumors (SEBAOUN, DELZANT and PEILLON, 1967). The fact that medroxyprogesterone acetate (Provera) may induce galactorrhea in 9% of cases after administration of a total of more than 1 g given in 10—20 injections, points to the possibility that it is partly metabolized to an estrogenic compound which affects the hypothalamus and the pituitary (BOLOGNESE, PIVER and FELDMAN, 1967).

j) Promotion of post-partum Lactation

LYONS, LI, AHMAD and RICE-WRAY (1968) injected 22 lactating women, most of whom complained of insufficient milk secretion, subcutaneously for one week with daily 10 mg doses of a human pituitary preparation which had about 5 IU/mg of pigeon crop-stimulating activity; about 1 IU/mg of somatotropin; and lactogenic activity detectable at the 20 µg dose in hypophysectomized rats. Placebo treatment for 1 week preceded or followed hormone injections. The suckling babies were weighed daily, and average daily weight changes were compared during the 2 regimens. In 16 cases the average weight gain during the hormone period was over 100% greater than that during the placebo treatment.

k) Gynecomastia

α) Morphology

Enlargement of the male breast, owing to an increase in the glandular component, is known as gynecomastia. The degree of enlargement is variable, ranging from a barely visible, small, central, subareolar disc of mammary tissue to an extension the size of a normal female adolescent breast. It may be unilateral or bilateral, frequently painful and tender, and, occasionally, emitting a secretion. Its presence is sometimes difficult to ascertain in obese males, since their breast enlargement may be due entirely, or in large part, to fat deposition (pseudo-gynecomastia).

The histopathology is characterized by stimulation of ducts and proliferation of stroma. The ducts undergo lengthening and branching, with budding and formation of new ducts without alveoli. Simultaneously, there is an increase in the bulk of the stromal tissue, which is often hyalinized. Stromal replacement of adipose tissue and periductal infiltrations with round cells are common. Since such changes may be induced by the administration of estrogen to males or to unstimulated females, it is logical to seek a similar explanation in patients with gynecomastia, but efforts in this direction have been only partially successful, especially because the condition is encountered in a great variety of apparently unrelated pathological disorders. Easier to explain is its occurrence in certain physiologic states.

β) Physiologic States

Neonatally, slight transitory breast enlargement is common in both sexes (witch milk). This is presumably due primarily to high levels of maternal estrogens. Gynecomastia, often slight, bilateral, and frequently painful, occurs in about two-thirds of boys during puberty (Froehlich Syndrome). It subsides spontaneously within 2 to 4 years in over 90 per cent of cases. Involutional enlargement occurs in some men later in life.

γ) *Pathological Conditions*

Superficially, these appear to be divisible into endocrine and non-endocrine disorders. It seems, however, that their origin is ultimately endocrine-conditioned. Chronic hyperestrogenism, possibly synergized by the action of one or more anterior pituitary hormones (i. e. prolactin or growth hormone), appears to be the principal etiologic factor. Estrogenic preponderance (without overt feminization) may arise through increased estrogen production, impaired estrogen inactivation in liver cirrhosis, and reduced androgen elaboration, or even increased androgens converted into estrogens. Augmented mammary sensitivity to circulating estrogens may explain certain innocuous familial cases, among others. B-avitaminosis may be a contributing factor, as vitamin B components are required for steroid metabolism in the liver. In most pathological conditions involving the endocrine glands absolute or relative hyperestrogenism is involved.

In general, gynecomastia appearing in, or after, the third decade of life is most often a manifestation of a serious underlying disease operating via hypothalamic release of prolactin. Examples are disorders of the testes, e. g. tumors, especially chorionepithelioma (HUNT and BUDD, 1939), hypogonadism (KLINEFELTER, REIFENSTEIN and ALBRIGHT, 1942), male pseudo-hermaphroditism, adrenal cortex tumors or hyperplasia (WILKINS, 1948), hypothyroidism (BERGSTRAND, 1955; SILVER, 1958) and pituitary tumors, e. g. acromegaly or chromophobe adenoma (DAVIDOFF, 1926). Other disorders, which possibly suppress the hypothalamic PIF, include hematochromatosis (S. C. FINCH and C. A. FINCH, 1955), liver cirrhosis due to alcohol abuse with impaired steroid metabolism and increased estrogen levels (EMERSON, 1966), malnutrition with vitamin B deficiency which also impairs estrogen cleavage (KLATSKIN, SALTER and HUMM, 1947), refeeding after starvation (PLATT, SCHULTZ and KUNSTADTER, 1946; JACOBS, 1948), bronchogenic carcinoma (DEL CASTILLO, DE LA BALZE and REFORZO MEMBRIVES, 1941), and traumatic paraplegia (KUPPERMAN, 1963 a, b). Obviously, gynecomastia readily develops after prolonged administration of estrogens for the treatment of prostate carcinoma, or of estrogen-contaminated drugs. It also occurs infrequently after treatment with chorionic gonadotropin, and steroid derivatives, as e. g. androgens (Lewin, 1941), digitalis (Lewinn, 1953), spironolactone (Mann, 1963; Sussman, 1963; Clark, 1965). Phenothiazines (Winnik and Tennenbaum, 1955; Margolis and Gross, 1967), isoniazid (INH) (Koang, Hu, Tchén and Chu, 1957; Molina, Aberkane and Conquydouard, 1957), ethionamide (Trecator) (Gernez-Rieux, Taquet and Maquet, 1963) and reserpine (Meyler, 1966) may also provoke gynecomastia via suppression of the hypothalamic PIF. Diabetes mellitus, generalized exfoliative dermatitis and many other miscellaneous conditions may sometimes be accompanied by gynecomastia. Here the hypothalamic relation is not clear.

Klinefelter's syndrome is characterized by gynecomastia with semineferous tubular hyalinization, small atrophic testes, elevated urinary FSH excretion, eunuchoidism and sometimes mental deficiency. It is due to the presence of a supernumerary chromosome—these patients having 47 chromosomes instead of the normal 46.

A prerequisite for treatment is thorough diagnostic investigation. Hormonal therapy is of no avail (norethandrolone up to 60 mg/d was proposed by Kupperman, 1963 a, b). Mammoplasty is indicated exclusively for cosmetic reasons and only when spontaneous regression has not taken place. This surgical procedure usually improves the patient's ability to carry on a normal social life as a male, preventing recurrent embarrassment due to the feminine breast development. There are, however, cases who end up as transvestites (Foss, 1958).

l) Conclusion

In lactating women endogenous prolactin suppresses FSH for 4 weeks and LH for 12 weeks. The knowledge of these terms is important for patients who wish to start the use of contraceptive pills without impairing lactation. If lactation is to be suppressed, stilbestrol should be given at doses of over 10 mg/d, combination with progestational or androgenic steroids being optional. Such treatment does not affect galactorrhea in the Chiari-Frommel syndrome or in the Ahumada-del Castillo syndrome. This fact points to our theory that normal lactation is produced by a prolactin-releasing factor sensitive to steroids, while pathological galactorrhea is produced by an impaired prolactin-inhibiting factor. These diseases do not react to steroid therapy; they are, however, slightly responsive to clomiphene.

Hypothalamic depressants, such as derivatives of phenothiazine, reserpine, butyrophenone, methyldopa, medroxyprogesterone acetate (in doses exceeding 1 g), can produce galactorrhea by suppressing the prolactin-inhibiting center. Recently, human growth hormone has been successfully applied in the promotion of post-partum lactation.

Gynecomastia in man is produced by a similar hypothalamic mechanism, mostly due to serious underlying disease, if not produced by drugs accidentally contaminated with estrogenic hormone or by stilbestrol phosphate used for the treatment of inoperable prostate carcinoma.

Chapter 29

Conclusions

The mammary gland represents a unique organ from a neuroendocrine point of view, since its development and secretory function are regulated by a harmonious concord of most of the endocrine glands and the central nervous system. Few other tissues in the body are influenced by so many endocrine, neural, and environmental factors. All of the anterior and posterior pituitary hormones may act on the mammary gland, either directly or through hormones of target organs, and all of them are controled by the hypothalamus. Parathormone, insulin, placental hormones, and perhaps relaxin may also influence mammary function. Since the survival of most mammalian species would not be possible without milk, which requires an immense supply of protein, fats, sugar and calcium, it is perhaps not surprising that so many mechanisms have been evoked to permit the growth and secretory function of the mammary gland.

The hormones mainly responsible for lactation are prolactin and growth hormone. Estrogen, progesterone, corticosteroids, thyroidine and insulin have been shown to be of secondary importance in mammary development. Prolactin and growth hormone are considered to be principally involved since they alone can induce mammary lobulo-alveolar development in the absence of gonadal steroids, whereas estrogen and progesterone have but little effect in the absence of the pituitary hormones. These two steroids are believed to act mainly by increasing the release of prolactin and growth hormone from the anterior pituitary, and by synergizing with these hormones.

Chapter 30

Research Problems

The goal of producing hypothalamic lactation has become more accessible but still requires much further study from varying aspects. Substances more specific than phenothiazine and its sulfoxides should in particular be synthetized. Reserpine and haloperidol derivatives are at present being studied. The combination of such prolactin releasers with STH releasers becomes imperative in cows and sheep, and we have now begun a large-scale study of the possibilities of releasing growth hormone in laboratory animals. Paracetamol and some of its derivatives prepared by us may release growth hormone (Dikstein, Grotto, Zor, Tamari and Sulman, 1966). Combinations will be studied in cows and sheep.

To test the hypothalamic prolactin releasers on cows does not seem very promising since apparently cows do not require prolactin for lactation. We, therefore, studied this method in sheep. The experiment was started in the 4th month of lactation when the daily milk yield had a tendency to decline. We tested the following eight compounds: piperacetazine (Quide), perphenazine sulfoxide (Schering), perphenazine (Trilafon), propericiazine (Neuleptil), reserpine (Serpasil), haloperidol (Halidol), trifluperidol (Trifludol), Sum-3170 (Wander). They were injected into 24 Awassi ewes either i. m. or i. v. for 6 days, in doses ranging between 0.02 and 0.5 mg daily. A slight increase in lactation was obtained only with propericiazine (0.5 mg i. m.), reserpine (0.05 mg i. m.) and haloperidol (0.1 mg i. v.). These experiments need further study and extension since the daily milk yield in sheep is subject to considerable fluctuations.

The major problem yet to be solved is elucidation of the structure of the hypothalamic factors responsible for lactation. It appears from our studies that phenothiazine derivatives and other psychopharmaca suppress the prolactin-inhibiting factor, whereas delivery and nursing release a prolactin-releasing factor. We feel, therefore, that, in addition to the prolactin-inhibiting factor of the hypothalamus, a prolactin-releasing factor should be studied. What facilitates this approach is that such a factor might possibly also stimulate other pituitary hormones, e. g. those required for preparation of lactation (FSH and LH), milk secretion (LTH and ACTH) and for maintenance of lactation (STH and TSH). Stimulation of a prolactin-releasing center may become feasible in farm animals through the stereotaxic approach. The methods we have begun using in laboratory animals are:

a) electrical stimulation of various centers that are responsible for lactation and maternal behavior;

b) removal of lactation-releasing factors from the hypothalamus of lactating animals through introducing capillaries which work by capillary attraction;

c) elution of lactation-releasing factors from suckling farm animals by intraventricular injections and perfusion of the cerebral ventricles, especially on the 10th day post-partum which has been found to be the peak period.

A technique not yet in use in domestic animals would be the injection of lactation-promoting drugs into the internal carotid artery, a site from which it would directly stimulate the hypothalamus and pituitary. This method could then be used for initiating lactation and increasing lactation to maximum levels for an extended period. The effect of oxytocin in this system requires further study.

Recently we found that perphenazine lactation in rats is impaired by adrenalectomy in spite of high blood prolactin levels. This would justify the re-examination of the amorphous steroid fraction of the adrenal cortex. It may contain a "corticolactin", differing from the hitherto identified corticosteroids.

It is most unlikely that hypothalamic milk releasers will be used per os in veterinary or in human medicine, since concentration of the releaser substance in the milk is just as high as in the blood. This has been confirmed by us by using radioactive phenothiazine derivatives in rabbits. Still, their use in farm animals may be possible by direct implantation into the hypothalamus, which would obviate any excretion either in the urine or the milk.

Modern methods of radioimmunoassay, which would permit measuring hormone levels at the different stages of pregnancy and lactation, could yield exact information on the gamut of the hormones participating in lactation at each stage. Thus, their practical use, either by injection or through release by chemical substances, would become feasible in the promotion of lactation.

References

Abood, L. G.: Effect of chlorpromazine on phosphorylation of brain mitochondria. Proc. Soc. exp. Biol. (N. Y.) **88,** 688 (1955).

Ahrén, K., Jacobsohn, D.: Mammary gland growth in hypophysectomized rats injected with ovarian hormones and insulin. Acta physiol. scand. **37,** 190 (1956).

— — The action of cortisone on the mammary glands of rats under various states of hormonal imbalance. Acta physiol. scand. **40,** 254 (1957).

Alloiteau, J. J.: Action inhibitrice de la chlorpromazine sur le déclenchement de deux types de pseudo-gestation chez la ratte. C. R. Soc. Biol. (Paris) **151,** 207 (1957).

Anderson, R. R., Turner, C. W.: Synergistic effect of various hormones on experimental mammary gland grown in mice. Proc. Soc. exp. Biol. (N. Y.) **113,** 308 (1963).

Andersson, B., McCann, S. M.: Drinking, antidiuresis and milk ejection from electrical stimulation within the hypothalamus of the goat. Acta physiol. scand. **35,** 191 (1955).

Apostolakis, M.: Lack of high growth hormone activity in highly active human pituitary prolactin extracts. In: Excerpta medica, international congress series no. 83, proceedings of the second international congress of endocrinology, London, August, 1964. Amsterdam: Excerpta medica foundation 1964, p. 1257.

Arai, Y., Lee, T. H.: A double-antibody radioimmunoassay procedure for ovine pituitary prolactin. Endocrinology **81,** 1040 (1967).

Argonz, J., Del Castillo, E. B.: A syndrome characterized by estrogenic insufficiency, galactorrhea and decreased urinary gonadotropin. J. clin. Endocr. **13,** 79 (1953).

Arimura, A., Saito, T., Müller, E. E., Bowers, C. Y., Sawano, S., Schally, A. V.: Absence of prolactin-release inhibiting activity in highly purified LH-releasing factor. Endocrinology **80,** 972 (1967).

Armstrong, D. T., Black, D. L.: Influence of luteinizing hormone on corpus luteum metabolism and progesterone biosynthesis throughout the bovine estrous cycle. Endocrinology **78,** 937 (1966).

— Kilpatrick, R., Greep, R. O.: *In vitro* and *in vivo* stimulation of glycolysis in prepubertal rat ovary by luteinizing hormone. Endocrinology **73,** 165 (1963).

— O'Brien, J., Greep, R. O.: Effect of luteinizing hormone on progestin biosynthesis in the luteinized rat ovary. Endocrinology **75,** 488 (1964).

ASSAEL, M., GABAI, F., WINNIK, H. Z., KHAZAN, N., SULMAN, F. G.: Recession of exophthalmos after massive doses of reserpine. Lancet **1960 I,** 499.

ASTWOOD, E. B.: The regulation of corpus-luteum function by hypophysial luteotrophin. Endocrinology **28,** 309 (1941).

— Tests for luteotrophins. Ciba Foundation Colloq. Endocrinology **5,** 74 (1953).

AVERILL, R. L. W.: Restoration of lactation in rats with hypothalamic lesions which inhibit lactation. J. Endocr. **31,** 191 (1965).

BAHN, R. C., BATES, R. W.: Histologic criteria for detection of prolactin: lack of prolactin in blood and urine of human subjects. J. clin. Endocr. **16,** 1337 (1956).

BAILEY, G. L., BARTLETT, S., FOLLEY, S. J.: Use of L-thyroxine by mouth for stimulating milk secretion in lactating cows. Nature (Lond.) **163,** 800 (1949).

BALDWIN, R. L., MARTIN, R. J.: Effects of hypophysectomy and several hormone replacement therapies upon patterns of nucleic acid and protein synthesis and enzyme levels in lactating rat mammary glands. J. Dairy Sci. **51,** 748 (1968 a).

— — Protein and nucleic acid synthesis in rat mammary glands during early lactation. Endocrinology **82,** 1209 (1968 b).

BAN, T., SHIMIZU, S., KUROTSU, J.: Experimental studies on the relationship between the hypothalamus including area preoptica and lactation in rabbits. Med. J. Osaka Univ. **8,** 345 (1958).

BARANYAI, P., NAGY, I., KURCZ, M.: Kühlbare horizontale Stärke-Gel-Elektrophorese-Einrichtung für die Fraktionierung der Eiweißstoffe in der Adenohypophyse. Clin. chim. Acta **16,** 449 (1967).

BARDIN, C. W., LIEBELT, A. G., LIEBELT, R. A.: The direct effect of pituitary isografts on mammary gland development in the mouse. Proc. Soc. exp. Biol. (N. Y.) **110,** 716 (1962).

BARLET, J. P.: Effets de la thyrocalcitonine sur la calcémie et la phosphatémie de la vache laitière. C. R. Acad. Sci. (Paris) Ser. D **265,** 1075 (1967).

BARNAWELL, E. B.: Analysis of the direct action of prolactin and steroids on mammary tissue of the dog in organ culture. Endocrinology **80,** 1083 (1967).

BARRACLOUGH, C. A.: Induction of pseudopregnancy in the rat by reserpine and chlorpromazine. Anat. Rec. **127,** 262 (1957).

— SAWYER, C. H.: Blockade of the release of pituitary ovulating hormone in the rat by chlorpromazine and reserpine: possible mechanisms of of action. Endocrinology **61,** 341 (1957).

— — Induction of pseudopregnancy in the rat by reserpine and chlorpromazine. Endocrinology **65,** 563 (1959).

BARRETT, R. J., FRIESEN, H., ASTWOOD, E. B.: Characterization of pituitary and peptide hormones by electrophoresis in starch gel. J. biol. Chem. **237,** 432 (1962).

Bartlett, S., Burt, A. W. A., Folley, S. J., Rowland, S. J.: Relative galactopoietic effects of 3:5:3'-triiodo-L-thyronine and L-thyroxine in lactating cows. J. Endocr. **10,** 192 (1954).

Bartosik, D., Romanoff, E. B., Watson, D. J., Scricco, E.: Luteotropic effects of prolactin in the bovine ovary. Endocrinology **81,** 186 (1967).

Bates, R. W., Garrison, M. M., Cornfield, J.: An improved bio-assay for prolactin using adult pigeons. Endocrinology **73,** 217 (1963).

— Lahr, E. L., Riddle, O.: The gross action of prolactin and follicle-stimulating hormone on the mature ovary and sex accessories of fowl. Amer. J. Physiol. **111,** 361 (1935).

— Riddle, O.: Effect of volume used for injection in micro-assay of prolactin. Proc. Soc. exp. Biol. (N. Y.) **44,** 505 (1940).

— — Annual variation in the response of crop-sacs and viscera of pigeons to prolactin. Endocrinology **29,** 702 (1941).

Bayle, J. D., Assenmacher, I.: Absence de stimulation du jabot du pigeon après autogreffe hypophysaire. C. R. Acad. Sci. (Paris) **261,** 5667 (1965).

Beck, J. C., Gonda, A., Hamid, M. A., Morgen, R. O., Rubenstein, D., McGarry, E. E.: Some metabolic changes induced by primate growth hormone and purified ovine prolactin. Metabolism **13,** Suppl. 1108 (1964 a).

— McGarry, E. E., Dawson, K. G., Gonda, A., Hamid, M. A., Rubinstein, D.: Comparative metabolic actions of primate growth hormone and non-primate prolactin in man. In: Excerpta medica, International congress series no. 83, proceedings of the second international congress of endocrinology, London, Aug., 1964. Amsterdam: Excerpta medica Foundation 1964 b, p. 1242.

Ben-David, M.: A sensitive bioassay for prolactin based on H^3-methyl-thymidine uptake by the pigeon-crop mucous epithelium. Proc. Soc. exp. Biol. (N. Y.) **125,** 705 (1967).

— Induction of hypothalamic lactation in Fischer inbred rats by depot injection of fluphenazine enanthate. Neuroendocrinology **3,** 65 (1968 a).

— The role of the ovaries in perphenazine-induced lactation. J. Endocr. **41,** 377 (1968 b).

— Mechanism of induction of mammary differentiation in Sprague-Dawley female rats by perphenazine. Endocrinology **83,** 1217 (1968 c).

— Chrambach, A.: Stimulation of pituitary release and production of prolactin by perphenazine. J. biol. Chem. (in press) (1970).

— Dikstein, S., Sulman, F. G.: Effects of different steroids on prolactin secretion in pituitary-hypothalamus organ co-culture. Proc. Soc. exp. Biol. (N. Y.) **117,** 511 (1964).

— — — Production of lactation by non-sedative phenothiazine derivatives. Proc. Soc. exp. Biol. (N. Y.) **118,** 265 (1965).

— — — Role of the thyroid in hypothalamic lactation. Proc. Soc. exp. Biol. (N. Y.) **121,** 873 (1966).

— Khazen, K., Khazan, N., Sulman, F. G.: Correlation between depressant and mammotropic effects of fifteen reserpine analogues. Arch. int. Pharmacodyn. **171,** 274 (1968).

BEN-DAVID, M., RODERIG, H., KHAZEN, K., SULMAN, F. G.: Effect of different steroids on lactating rats. Proc. Soc. exp. Biol. (N. Y.) **120**, 620 (1965).

— SULMAN, G. F.: The mechanism of systemic deciduoma production. Arch. int. Pharmacodyn. **159**, 161 (1966).

— — The role of the adrenals in perphenazine-induced lactation. Acta endocr. (Kbh.) (in press) (1970).

BENSON, G. K.: Pituitary-adrenal relationships in the retardation of mammary gland involution by oxytocin. J. Endocr. **20**, 91 (1960).

— COWIE, A. T., COX, C. P., GOLDZWEIG, S. A.: Effects of estrone and progesterone on mammary development in the guinea-pig. J. Endocr. **15**, 126 (1957).

— — TINDAL, J. S.: The pituitary and the maintenance of milk secretion. Proc. roy. Soc. B. **149**, 330 (1958).

— FOLLEY, S. J.: Oxytocin as stimulator for the release of prolactin from the anterior pituitary. Nature (Lond.) **177**, 700 (1956).

— — The effect of oxytocin on mammary gland involution in the rat. J. Endocr. **16**, 189 (1957).

BERGENSTAL, D. M., LIPSETT, M. B.: The anabolic effect of sheep prolactin in man. J. clin. Invest. **37**, 877 (1958).

BERGSTRAND, C. G.: A case of hypothyroidism with signs of precocious sexual development. Acta endocr. (Kbh.) **20**, 338 (1955).

BERNTHSEN, A.: Zur Kenntnis des Methylenblau und verwandter Farbstoffe. Ber. dtsch. chem. Ges. **16**, 2897 (1883).

BERSON, S. A., YALOW, R. S.: Immunoassay of protein hormones. In: The Hormones, Vol. IV. Eds.: G. PINCUS, K. V. THIMANN, and E. B. ASTWOOD. New York-London: Academic Press 1964, p. 557.

BERSWORDT-WALLRABE, I. VON, HERLYN, U., JANTZEN, K.: A modification of the pigeon crop sac assay for determination of lactogenic hormone (prolactin) in human serum. Acta endocr. (Kbh). suppl. **100**, 169 (1965).

BHARGAVA, K. P., CHANDRA, O.: Tranquillizing and hypotensive activities of twelve phenothiazines. Brit. J. Pharmacol. **22**, 154 (1964).

BISSET, G. W.: The milk-ejection reflex and actions of oxytocin, vasopressin and synthetic analogues on the mammary gland. In: Handbuch der experimentellen Pharmakologie. Vol. XXIII; Neurohypophysial hormones and similar polypeptides. Berlin-Heidelberg-New York: Springer 1968, p. 475.

BLAXTER, K. L.: The preparation and biological effects of iodinated proteins. The effect of iodinated protein feeding on the lactating cow. I. The effects of preparations of low activity and of iodinated ardein. J. Endocr. **4**, 237 (1945).

BOLOGNESE, R. J., PIVER, M. S., FELDMAN, J. D.: Galactorrhea and abnormal menses associated with a long-acting progesterone. J. Amer. med. Ass. **199**, 42 (1967).

BOOT, L. M., ROEPCKE, G., MUEHLBOCK, O.: Prolacting-producing pituitary tumors arising in pituitary isografts in mice. In: Excerpta Medica, International congress series No. 33, Proceedings of the second international congress of endocrinology, London, August 1964, 1058. Amsterdam: Excerpta medica foundation 1964.

BRAUDE, R., MITCHELL, K. G.: Observations on the relationship between oxytocin and adrenaline in milk ejection in the sow. J. Endocr. **8,** 238 (1952).

BROWN-GRANT, K., CROSS, B. A. (Eds.): Recent studies on the hypothalamus. Brit. med. Bull. **22,** 195 (1966).

BRUCE, H. M.: Suckling stimulus and lactation. Proc. roy. Soc. B **149,** 421 (1958).

— PARKES, A. S.: An olfactory block to nidation in mice. In: Advances in neuroendocrinology, p. 282. Ed.: A. V. NALBANDOV. Urbana (Ill.): Univ of Illinois Press 1963.

— SLOVITER, H. A.: Effect of destruction of thyroid tissue by radioactive iodine on reproduction in mice. J. Endocr. **15,** 72 (1957).

BRUMBY, P. J., HANCOCK, J.: The galactopoietic rate of growth hormone in dairy cattle. N. Z. Jl. Sci. Technol. **36,** 417 (1955).

BRYANT, G. D., CONNAN, R. H., GREENWOOD, F. C.: Changes in plasma prolactin induced by acepromazine in sheep. J. Endocr. **41,** 613 (1968).

— GREENWOOD, F. C.: Plasma prolactin levels in the goat; physiological and experimental modification. J. Endocr. **40,** IV (1968).

— TECH, B., GREENWOOD, F. C., LINZEL, J. L.: Radioimmunoassay of ovine, caprine and bovine prolactin in plasma; physiological and experimental changes in levels. Excerpta Medica, International congress series No. 157, Third international congress of endocrinology, Mexico, D. F., June 30—July 5, 1968. Amsterdam: Excerpta medica foundation 1968, p. 76.

BUTT, W. R.: In: Immunochemical studies on human growth hormone; a consideration of the human growth hormone-antihuman growth hormone system and its application to the assay of growth hormone. Eds.: M. M. and S. L. KAPLAN: Questions in Grumbach. Ciba Found. Colloq. Endocr. **14,** 103 (1962).

CANFIELD, C. J., BATES, R. W.: Nonpuerperal galactorrhea. New Engl. J. Med. **273,** 897 (1965).

CASTILLO, E. B. DEL, DE LA BALZE, F. A., REFORZO MEMBRIVES, J.: Ginecomastia y cancer de pulmón. Sem. méd. (B. Aires) **1,** 1419 (1941).

CHADWICK, A., FOLLEY, S. J., GEMZELL, C. A.: Lactogenic activity of human pituitary growth hormone. Lancet **1961 I,** 241.

CHEN, T. T., JOHNSON, R. E., LYONS, W. R., LI, C. H., COLE, R. D.: Hormonally induced mammary growth and lactation in the absence of the thyroid. Endocrinology **57,** 153 (1955).

CHEN, C. L., MEITES, J.: Effect of thyroxine and thiouracil on hypothalamic PIF and pituitary prolactin levels. Proc. Soc. exp. Biol. (N. Y.) **131,** 576 (1969).

CHEN, H. C., WILHELMI, A. G.: Purification of human growth hormone. Progr. 46th meet. Endocrine Soc. **46,** 115 (1964).

CHIARI, J. B. V. L., BRAUN, C., SPAETH, J.: Klinik der Geburtshilfe und Gynaekologie. Erlangen: Enke 1852.

CHILD, K. J., SUTHERLAND, P., TOMICH, E. G.: Some effects of reserpine on barbitone anaesthesia in mice. Biochem. Pharmacol. **6**, 252 (1961).

— — — Barbitone induction time, body temperature and the actions of reserpine on mice. Biochem. Pharmacol. **11**, 475 (1962).

CHOUDARY, J. P., GREENWALD, G. S.: Effect of an ectopic pituitary gland on luteal maintenance in the hamster. Endocrinology **81**, 542 (1967).

CHRAMBACH, A., BEN-DAVID, M.: The molecular identity of rat prolactin derived from serum and pituitary under conditions of both natural and pharmacological-induced lactation. J. biol. Chem. (in press) (1970).

CLARK, E.: Spironolactone therapy and gynecomastia. J. Amer. med. Ass. **193**, 163 (1965).

CLEMENS, J. A., MEITES, J.: Prolactin implant into the median eminence inhibits pituitary prolactin secretion, mammary growth and luteal function. Physiologist **10**, 144 (1967).

— — Inhibition by hypothalamic prolactin implants of prolactin secretion, mammary growth and luteal function. Endocrinology **82**, 878 (1968).

— MINAGUCHI, H., STOREY, R., VOOGT, J. L., MEITES, J.: Induction of precocious puberty in female rats by prolactin. Neuroendocrinology **4**, 150 (1969).

— SAR, M., MEITES, J.: Termination of pregnancy by prolactin implants in the median eminence of the rat. Proc. Int. Un. Physiol. Sci. **7**, 1 (1968).

— — — Termination of pregnancy in rats by a prolactin implant in median eminence. Proc. Soc. exp. Biol. (N. Y.) **130**, 628 (1969).

CLEVERLEY, J. D.: The detection of oxytocin release in response to conditioned stimuli associated with machine milking in the cow. J. Endocr. **40**, II (1967).

COHERE, G., BOUSQUET, J., MEUNIER, J. M.: Ultrastructure d'explants de glande pituitaire de ratte adulte cultivés sur milieux artificiels; action de variations hormonales. C.R. Soc. Biol. (Paris) **158**, 1056 (1964).

CONTOPOULOS, A. N., SIMPSON, M. E., KONEFF, A.: Pituitary function in the thyroidectomized rat. Endocrinology **63**, 642 (1958).

COPPEDGE, R. L., SEGALOFF, A.: Urinary prolactin excretion in man. J. clin. Endocr. **11**, 465 (1951).

COPPOLA, J. A., LEONARDI, R. G., LIPPMANN, W., PERRINE, J. W., RINGLER, I.: Induction of pseudopregnancy in rats by depletors of endogenous catecholamines. Endocrinology **77**, 485 (1965).

CORBIN, A., COHEN, A. I.: Effect of median eminence implants of LH on pituitary LH of female rats. Endocrinology **78**, 41 (1966).

— STORY, J. C.: "Internal" feedback mechanism: Response of pituitary FSH and of stalk-median eminence follicle stimulating hormone-releasing factor to median eminence implants of FSH. Endocrinology **80**, 1006 (1967).

CORDELLI, F., CASTELLI, A., GORELLI, L.: Effect of hormones on lactation: I. Effect of Insulin. Lattante **31**, 546 (1960). (Title in Italian: Azioni ormoniche sulla lattazione Nota I. Azione dell'insulina.)

COTES, P. M., CRICHTON, J. A., FOLLEY, S. J., YOUNG, F. G.: Galactopoietic activity of purified anterior pituitary growth hormone. Nature (Lond.) **164**, 992 (1949).

COURVOISIER, S., DUCROT, R., JULOU, L.: Nouveaux aspects expérimentaux de l'activité centrale des dérivés de la phénothiazine. In: Psychotropic drugs, p. 373. Eds.: S. GARATTINI and V. GHETTI. Amsterdam: Elsevier 1957.

— FOURNEL, J., DUCROT, R., KOLSKY, M., KOETSCHET, P.: Propriétés pharmacodynamiques du chlorhydrate de chloro-3 (diméthyl-amino-3 propyl)-10-phénothiazine (4.560 R.P.); étude expérimentale d'un nouveau corps utilisé dans l'anesthésie potentialisée et dans l'hibernation artificielle. Arch. int. Pharmacodyn. **92**, 305 (1953).

COWIE, A. T.: Influence on the replacement value of some adrenal-cortex steroids of dietary sodium and synergism of steroids in lactating adrenalectomized rats. Endocrinology **51**, 217 (1952).

— Hormones and lactation. N.I.R.D. paper No. 1762. Nat. Agr. Adv. Connc. Quart. Rev. **2**, 32 (1956).

— The maintenance of lactation in the rat after hypophysectomy. J. Endocr. **16**, 135 (1957).

— The hormonal control of milk secretion. In: Milk; the mammary gland and its secretion. Vol. I, p. 163. Eds.: S. K. KON and A. T. COWIE. New York-London: Academic Press 1961.

— Anterior pituitary function in lactation. In: The Pituitary Gland. Vol. II, p. 412. Eds.: G. W. HARRIS and B. T. DONOVAN. London: Butterworths 1966.

— FOLLEY, S. J.: The role of the adrenal cortex in mammary development and its relation to the mammogenic action of the anterior pituitary. Endocrinology **40**, 274 (1947).

— — Neurophysial hormones and the mammary gland. In: The neurohypophysis, p. 183. Ed.: H. HELLER. London: Butterworths 1957.

— — The mammary gland and lactation. In: Sex and internal secretions. Vol. I, p. 590. Ed.: W. C. YOUNG, 3rd ed. Baltimore: Williams & Wilkins 1961.

— — MALPRESS, F. H., RICHARDSON, K. C.: Studies on the hormonal induction of mammary growth and lactation in the goat. J. Endocr. **8**, 64 (1952).

— HARTMANN, P. E., TURVEY, A.: The maintenance of lactation in the rabbit after hypophysectomy. J. Endocr. **43**, 651 (1969).

— KNAGGS, G. S., TINDAL, J. S., TURVEY, A.: The milking stimulus and mammary growth in the goat. J. Endocr. **40**, 243 (1968).

— TINDAL, J. S.: Maintenance of lactation in adrenalectomized rats with aldosterone and 9 halo derivatives of hydrocortisone. Endocrinology **56**, 612 (1955).

COWIE, A. T., TINDAL, J. S.: Adrenalectomy in the goat; replacement therapy and the maintenance of lactation. J. Endocr. **16,** 403 (1958).
— — BENSON, G. K.: Pituitary grafts and milk secretion in hypophysectomized rats. J. Endocr. **21,** 115 (1960).
COX, W. M., MUELLER, A. J.: The composition of milk from stock rats and an apparatus for milking small laboratory animals. J. Nutr. **13,** 249 (1937).
CRONIN, T. J.: Influence of lactation upon ovulation. Lancet **1968 II,** 422.
CROSS, B. A.: Sympathetico-adrenal inhibition of the neurohypophysical milk ejection mechanism. J. Endocr. **9,** 7 (1953).
— Neurohormonal mechanisms in emotional inhibition of milk ejection. J. Endocr. **12,** 29 (1955).
— Neural control of lactation. In: Milk, the mammary gland and its secretion, Vol. I. Eds.: S. K. KON and A. T. COWIE. New York-London: Academic Press 1961, p. 229.
— HARRIS, G. W.: The role of the neurohypophysis in the milk-ejection reflex. J. Endocr. **8,** 148 (1952).
DANON, A., DIKSTEIN, S., SULMAN, F. G.: Stimulation of prolactin secretion by perphenazine in pituitary-hypothalamus organ culture. Proc. Soc. exp. Biol. (N. Y.) **114,** 366 (1963).
— — — In-vivo assay for hypothalamic prolactin-inhibiting factor. J. Endocr. **46,** 237 (1970).
— SULMAN, F. G.: Storage of prolactin-inhibiting factor in the hypothalamus of perphenazine-treated rats. Neuroendocrinology (in press) (1970).
DAO, T. L., GAWLAK, D.: Direct mammotrophic effect of a pituitary homograft in rats. Endocrinology **72,** 884 (1963).
DAVIDOFF, L. M.: Studies in acromegaly. III. The anamnesis and symptomatology in one hundred cases. Endocrinology **10,** 461 (1926).
DE FEO, V. J.: Effect of large doses of reserpine on the deciduoma response. Anat. Rec. **127,** 409 (1957).
— REYNOLDS, S. R. M.: Modification of the menstrual cycle in rhesus monkeys by reserpine. Science **124,** 726 (1956).
DELAY, J., DENIKER, P.: Les hormones. Journées thérap. Paris **1,** 97 (1953).
DE LEE, J. B.: The principles and practice of obstetrics. 4th ed. Philadelphia: Saunders 1924.
DESCLIN, L.: A propos du mécanisme des oestrogènes sur la lobe antérieur de l'hypophyse. Ann. Endocr. (Paris) **11,** 656 (1950).
— Hypothalamus et libération d'hormone lutéotrophique. Expériences de greffe hypophysaire chez le rat hypophysectomisé. Action lutéotrophique de l'ocytocine. Ann. Endocr. (Paris) **17,** 586 (1956).
— Action de la réserpine sur la glande mammaire et sur l'ovaire du rat. Ann. Endocr. (Paris) **19,** 1225 (1958).
— FLAMENT-DURAND, J.: Effect of reserpine on the morphology of pituitaires grafted into the hypothalamus or under the kidney capsule in the rat. J. Endocrinol. **43,** LIX (1969).

DESJARDINS, C., PAAPE, N. J., TUCKER, H. A.: Contribution of pregnancy, fetal placentas and deciduomas to mammary gland and uterine development. Endocrinology **83**, 907 (1968).

DHARIWAL, A. P. S., GROSVENOR, C. E., ANTUNES-RODRIGUES, J., MCCANN, S. M.: Correspondence of prolactin-inhibition and LH-releasing activities after gel filtration of hypothalamic extracts on sephadex. Progr. 47th meet. Endocrine Soc. **47**, 127 (1965).

— — — — Studies on the purification of ovine prolactin-inhibiting factor. Endocrinology **82**, 1236 (1968).

DIKSTEIN, S., GROTTO, M., ZOR, U., TAMARI, M., SULMAN, F. G.: The stimulatory effect of paracetamol and its derivatives on growth and the rat tibia test. J. Endocr. **36**, 257 (1966).

— SULMAN, F. G.: Drugs acting on the hypothalamus-anterior pituitary axis. Obstet. gynec. Surv. **21**, 531 (1966).

DILLEY, W. G., NANDI, S.: Rat mammary gland differentiation *in vitro* in the absence of steroids. Science **161**, 59 (1968).

DIXON, J. S., LI, C. H.: Chemistry of prolactin. Metabolism **13**, 1093 (1964).

DJOJOSOEBAGIO, S., TURNER, C. W.: Effect of a combination of lactogenic, growth, thyroid, and parathyroid hormones on lactation in rats. Proc. Soc. exp. Biol. (N. Y.) **116**, 213 (1964).

DOMANSKI, E., DOBROWOLSKI, W.: Perfusion of an organ in situ as a method in endocrinological investigations. In: Excerpta medica, international congress series no. 111, second international congress on hormonal steroids, Milan, Italy, May 23—28, 1966. Amsterdam: Excerpta medica foundation 1966, p. 259.

DONOVAN, B. T.: The effect of pituitary stalk section on luteal function in the ferret. J. Endocr. **27**, 201 (1963 a).

— Pituitary stalk section and corpus luteum function in the ferret. J. Physiol. (Lond.) **165**, 23P (1963 b).

— VAN DER WERFF TEN BOSCH, J. J.: The hypothalamus and lactation in the rabbit. J. Physiol. (Lond.) **137**, 410 (1957).

DRESEL, I.: The effect of prolactin on the estrus cycle of non-parous mice. Science **82**, 173 (1935).

DURLACH, J.: L'action mammotrope de la réserpine. Presse méd. **65**, 2060 (1957).

EBERT, A. G., HESS, S. M.: The distribution and metabolism of fluphenazine enanthate. J. Pharmacol. exp. Ther. **148**, 412 (1965).

ECKSTEIN, B.: The development of the ovary of the rat. II. The effect of chorionic gonadotrophin and serum gonadotrophin. Acta endocr. (Kbh.) **41**, 35 (1962).

EHRHARDT, K.: Über das Laktationshormon des Hypophysenvorderlappens. Münch. med. Wschr. **29**, 1163 (1936).

ELLIS, S., GRINDELAND, R. E., NUENKE, J. M., CALLAHAN, P. X.: Purification and properties of rat prolactin. Endocrinology **85**, 886 (1969).

ELY, F., PETERSON, W. E.: Factors involved in the ejection of milk. J. Dairy Sci. **24**, 211 (1941).

EMERSON, K., JR.: Diseases of the breast. In: Principles of internal medicine, p. 532. Eds.: T. R. HARRISON, R. D. ADAMS, I. L. BENNETT, W. H. RESNIK, G. W. THORN and M. M. WINTROBE. New York: McGraw Hill 1966.

EVANS, H. M., SIMPSON, M. E., LYONS, W. R., TURPEINER, K.: Anterior pituitary hormones which favor the production of traumatic uterine placentomata. Endocrinology **28,** 933 (1941).

EVERETT, J. W.: Evidence suggesting a role of the lactogenic hormone in the estrous cycle of the albino rat. Endocrinology **35,** 507 (1944).

— Luteotrophic function of autografts of the rat hypophysis. Endocrinology **54,** 685 (1954).

— Functional corpora lutea maintained for months by autografts of rat hypophyses. Endocrinology **58,** 786 (1956).

— Central neural control of reproductive functions of the adenohypophysis. Physiol. Rev. **44,** 373 (1964).

— QUINN, D. L.: Differential hypothalamic mechanisms inciting ovulation and pseudopregnancy in the rat. Endocrinology **78,** 141 (1966).

FAURET, E.: Vergleichende Untersuchungen über die Entwicklung und Funktion der Milchdrüsen. Experimentelle Untersuchungen über die Wirkung von Follikelhormonzufuhr auf die Milchdrüsen säugender Ratten. Arch. Gynäk. **171,** 342 (1941).

FELDBERG, W.: A new concept of temperature control in the hypothalamus. Proc. roy. Soc. Med. **58,** 395 (1965).

FERGUSON, K. A., WALLACE, A. L. C.: Starch-gel electrophoresis of anterior pituitary hormones. Nature (Lond.) **190,** 629 (1961).

FINCH, S. C., FINCH, C. A.: Idiopathic hemachromatosis, an iron storage disease. Medicine (Baltimore) **34,** 381 (1955).

FISKIO, P. W.: Combined steroids to suppress lactation. GP (Kansas) **11,** 70 (1955).

FLERKO, B.: Control of gonadotropin secretion in the female. In: Neuroendocrinology. Vol. I, p. 613. Eds.: L. MARTINI and W. F. GANONG. New York-London: Academic Press 1966.

— SZENTAGOTHAI, J.: Estrogen sensitive nervous structures in the hypothalamus. Acta endocr. (Kbh.) **26,** 121 (1957).

FLORINI, J. R., TONELLI, G., BREUER, C. B., COPPOLA, J., RINGLER, I., BELL, P. H.: Characterization and biological effects of purified placental protein (human). Endocrinology **79,** 692 (1966).

FLUECKIGER, E., WAGNER, H. R.: 2-Br-Alpha-Ergokryptin: Beeinflussung von Fertilität und Laktation bei der Ratte. Experientia (Basel) **24,** 1130 (1968).

FLUX, D. S.: The effect of adrenal steroids on the growth of the mammary glands, uteri, thymus and adrenal glands of intact, ovariectomized and estrone-treated ovariectomized mice. J. Endocr. **11,** 238 (1954).

— The value of some steroids in replacement therapy in adrenalectomized and adrenalectomized-ovariectomized lactating rats J. Endocr. **12,** 57 (1955).

FLUX, D. S., FOLLEY, S. J., ROWLAND, S. J.: The effect of adrenocorticotrophic hormone on the yield and composition of the milk of the cow. J. Endocr. **10**, 333 (1954).
— MUNFORD, R. E.: The effect of adrenocorticotrophin on the mammary glands of intact mature female mice of the CHI strain. J. Endocr. **14**, 343 (1957).
FOLLEY, S. J.: The adrenal cortex and the mammary gland. In: The Suprarenal Cortex. Ed.: J. M. YOFFEY. Proceedings of the Colston Research Society's Fifth Symposium. Bristol, April 1952. London: Butterworth 1952 a.
— Lactation. In: Marshall's physiology of reproduction. Vol. II, p. 525. Ed.: A. S. PARKES. London: Longmans, Green 1952 b.
— The physiology and biochemistry of lactation. Edinburgh-London: Oliver and Boyd 1956.
— COWIE, A. T.: Adrenalectomy and replacement therapy in lactating rats. Yale J. Biol. Med. **17**, 67 (1944).
— KON, S. K.: The effect of sex hormones on lactation in the rat. Proc. roy. Soc. B **124**, 476 (1938).
— MALPRESS, F. H.: Hormonal control of mammary growth. In: The Hormones. Vol. I, p. 695. Eds.: G. PINCUS and K. V. THIMANN. New York: Academic Press 1948 a.
— — Hormonal control of lactation. In: The Hormones. Vol. I, p. 745. Eds.: G. PINCUS and K. V. THIMANN. New York: Academic Press 1948 b.
— YOUNG, F. G.: Further experiments on the continued treatment of lactating cows with anterior pituitary extracts. J. Endocr. **2**, 226 (1940).
FORBES, A. P., HENNEMAN, P. H. GRISWOLD, G. C., ALBRIGHT, F.: A syndrome distinct from acromegaly, characterized by spontaneous lactation, amenorrhea and low follicle-stimulating hormone excretion. J. clin. Endocr. **11**, 749 (1951).
— — — — Syndrome characterized by galactorrhea, amenorrhea, and low urinary FSH: Comparison with acromegaly and normal lactation. J. clin. Endocr. **14**, 265 (1954).
FORREST, J. S., PIETTE, L. H.: New techniques of identification of individual oxidative metabolites derived from phenothiazine and related compounds. Neuropsychopharmacol. **3**, 397 (1964).
FORSYTH, I. A.: Lactogenic and pigeon crop-stimulating activities of a human placental lactogen preparation. J. Endocr. **37**, XXXV (1967).
— The lactogenic activity of human growth hormone *in vivo*. In: Excerpta media, international congress series no. 158, Growth hormone, Proceedings of the first international symposium, Milan, Italy, September, 1967. Amsterdam: Excerpta medica foundation 1968, p. 364.
— FOLLEY, S. J., CHADWICK, A.: Lactogenic and pigeon crop-stimulating activities of human pituitary growth hormone preparations. J. Endocr. **31**, 115 (1965).
FOSS, G. L.: Disturbances of Lactation. Clin. Obstet. Gynec. **1**, 245 (1958).

FRIESEN, H.: Purification of a placental factor with immunological and chemical similarity to human growth hormone. Endocrinology **76**, 369 (1965).

FROMMEL, R.: Über puerperale Atrophie des Uterus. Z. Geburtsh. Gynäk. **7**, 305 (1882).

— Bericht über Frauenkrankheiten der Städtischen Frauenklinik Wien. Z. Geburtsh. Gynäk. 1845—1855.

GALA, R. R., REECE, R. P.: Effect of a reserpine analogue, methyl-18-epi-o-methyl-reserpate hydrochloride, on lactogen release from rat anterior pituitary. Endocrinology **72**, 649 (1963).

— — Influence of hypothalamic fragments and extracts on lactogen production *in vitro*. Proc. Soc. exp. Biol. (N. Y.) **117**, 833 (1964 a).

— — Influence of estrogen on anterior pituitary lactogen production *in vitro*. Proc. Soc. exp. Biol. (N. Y.) **115**, 1030 (1964 a).

— — Influence of neurohumors on anterior pituitary lactogen production *in vitro*. Proc. Soc. exp. Biol. (N. Y.) **120**, 220 (1965 a).

— — *In vitro* lactogen production by anterior pituitaries from various species. Proc. Soc. exp. Biol. (N. Y.) **120**, 263 (1965 b).

GATSCHEW, E.: Reserpin, Prolactin-Freisetzung und Milchzuckersynthese. Arzneimittel-Forsch. **16**, 75 (1966).

GAUNT, R., CHART, J. J., RENZI, A. A.: Interactions of drugs with endocrines. Ann. Rev. Pharmacol. **3**, 109 (1963).

— EVERSOLE, W. J., KENDALL, E. C.: Influence of some steroid hormones on lactation in adrenalectomized rats. Endocrinology **31**, 84 (1942).

— RENZI, A. A., ANTONCHAK, N., MILLER, G. J., GILMAN, M.: Endocrine aspects of the pharmacology of reserpine. Ann. N. Y. Acad. Sci. **59**, 22 (1954).

GERNEZ-RIEUX, C., TACQUET, A., MAQUET, V.: Les perfusions d'éthionamide dans le traitement de la tuberculose pulmonaire. G. ital. Chemioter. **10**, 87 (1963).

GESCHICKTER, C. F.: Diseases of the breast. Philadelphia: Lippincott 1960.

GOSPODAROWICZ, D., LEGAULT-DEMARE, J.: Etude de l'activité biologique *in vitro* des hormones gonadotropes. IV. Action synergique de la prolactine et de l'hormone chorionique humaine sur le corps jaune de rat *in vitro*. Acta endocr. (Kbh.) **42**, 509 (1963).

GRAHAM, W. R., JR.: The action of thyroxine on the milk and milk-fat production of cows. Biochem. J. **28**, 1368 (1934).

GREENBLATT, R. B., MAHESH, V. B.: Induction of ovulation with clomiphene citrate. In: Yearbook of endocrinology, p. 248. Ed.: T. B. SCHWARTZ. Chicago: Yearbook medical publishers 1964—1965.

GREENWALD, G. S.: Luteotropic complex of the hamster. Endocrinology **80**, 118 (1967).

GREEP, R. O.: Physiology of the anterior hypophysis in relation to reproduction. In: Sex and internal secretions, Vol. I. 3rd ed. Ed.: W. C. YOUNG. Baltimore: Williams & Wilkins 1961, p. 240.

Greig, M. E., Mayberry, T. C.: The relationship between cholinesterase activity and brain permeability. J. Pharmacol. exp. Ther. **102,** 1 (1951).

Griffith, L. S., McBride, W. P. L.: Clinical uses of endometrial biopsy. J. Mich. med. Soc. **38,** 1064 (1939).

Grimm, E. G.: Non-puerperal galactorrhea. Quart. Bull. Northw. Univ. med. Sch. **29,** 350 (1955).

Grönroos, M., Kalliomaki, J. L., Keyriläinen, T. O., Marjanen, P.: Effects of reserpine and chlorpromazine on the mammary glands of the rat. Acta endocr. (Kbh.) **31,** 154 (1959).

Grosvenor, C. E.: Effect of experimentally induced hypo- and hyperthyroid states upon pituitary lactogenic hormone concentration in rats. Endocrinology **69,** 1092 (1961).

— Evidence that exteroceptive stimuli can release prolactin from the pituitary gland of the lactating rat. Endocrinology **76,** 340 (1965 a).

— Effect of nursing and stress upon prolactin-inhibiting activity of the rat hypothalamus. Endocrinology **77,** 1037 (1965 b).

— Effect of nursing and stress upon "prolactin-inhibiting" activity of rat hypothalamus. Fed. Proc. **24,** 190 (1965 c).

— Krulich, L., McCann, S. M.: Depletion of pituitary concentration of growth hormone as a result of suckling in the lactating rat. Endocrinology **82,** 617 (1968).

— McCann, S. M., Nallar, R.: Inhibition of nursing-induced and stress-induced fall in pituitary prolactin concentration in lactating rats by injection of acid extracts of bovine hypothalamus. Endocrinology **76,** 883 (1965).

— Turner, C. W.: Pituitary lactogenic hormone concentration and milk secretion in lactating rats. Endocrinology **63,** 535 (1958 a).

— — Effect of lactation upon thyroid secretion rate in the rat. Proc. Soc. exp. Biol. (N. Y.) **99,** 517 (1958 b).

— — Effect of growth hormone and oxytocin upon milk yield in the lactating rat. Proc. Soc. exp. Biol. (N. Y.) **100,** 158 (1959 a).

— — Thyroid hormone and lactation in the rat. Proc. Soc. exp. Biol. (N. Y.) **100,** 162 (1959 b).

— — Pituitary lactogenic hormone concentration during pregnancy in the rat. Endocrinology **66,** 96 (1960).

Guillemin, R.: Personal communication (1967).

— Hearn, W. R., Cheek, W. R., Housholder, D. E.: Control of corticotrophin release: further studies with *in vitro* methods. Endocrinology **60,** 488 (1957).

— Rosenberg, B.: Humoral hypothalamic control of anterior pituitary: A study with combined tissue cultures. Endocrinology **57,** 599 (1955).

— Sakiz, E.: Quantitative study of the response to LH after hypophysectomy in the ovarian ascorbic acid depletion test. Endocrinology **72,** 813 (1963).

Guth, P. S., Spirtes, M. A.: The phenothiazine tranquilizers: biochemical and biophysical actions. Int. Rev. Neurobiol. **7,** 231 (1964).

HAASE, H. J., JANSSEN, P. A. J.: The action of neuroleptic drugs. Chicago: Yearbook Medical Publ. 1965.

HAHN, D. W., TURNER, C. W.: Effects of ovariectomy-hypophysectomy and adrenalectomy-ovariectomy-hypophysectomy on feed intake and mammary gland growth as measured by DNA. Proc. Soc. exp. Biol. (N. Y.) **126,** 476 (1967).

HAMBERGER, L., AHRÉN, K.: Influence of the adrenal cortex on growth processes in the rat mammary gland. J. Endocr. **30,** 171 (1964).

HARRIS, G. W.: Neural control of the pituitary gland. London: Arnold 1955.

— DONOVAN, B. T., Eds.: The pituitary gland. Vol. III. London: Butterworths 1966.

HEITZMAN, R. J.: The hormonal induction of glucose-6-phosphate metabolizing enzymes in the mammary gland of the rabbit. J. Endocr. **41,** XVI (1968).

HERLYN, U., GELLER, H. F., VON BERSWORDT-WALLRABE, I., VON BERSWORDT-WALLRABE, R.: Pituitary lactogenic hormone release during onset of pseudopregnancy in intact rats. Acta endocr. (Kbh.) **48,** 220 (1965).

— JANTZEN, K., FLASKAMP, D., HOFFMANN, H., VON BERSWORDT-WALLRABE, I.: A modification of the pigeon crop sac assay for lactotrophic hormone determinations by means of the addition of prednisolone. Acta endocr. (Kbh.) **60,** 555 (1969).

HERMAN, H. A., GRAHAM, W. R., JR., TURNER, C. W.: The effect of thyroid and thyroxine on milk secretion in dairy cattle. Res. Bull. Mo. agric. Exp. Stn. No. **275** (1938).

HESSELTINE, H. C., BUSTAMANTE, J., NAVORI, C. A.: Limited influence of diethylstilbestrol on lactation. Amer. J. Obstet. Gynec. **69,** 686 (1955).

HILLIARD, J., SAWYER, C. H.: Effect of prolactin on steroidogenesis and cholesterol storage in the rabbit ovary. In: Excerpta medica, international congress series no. 111, second international congress on hormonal steroids, Milan, Italy, May 23—28, 1966, p. 195. Amsterdam: Excerpta medica foundation 1966.

— SPIES, H. G., LUCAS, L., SAWYER, C. H.: Effect of prolactin on progestin release and cholesterol storage by rabbit ovarian interstitium. Endocrinology **82,** 122 (1968).

HISAW, F. L.: The placental gonadotrophin and luteal function in monkeys (Macaca mulatta). Yale J. Biol. Med. **17,** 119 (1944).

HOLLISTER, L. E.: Adverse reactions to phenothiazines. J. Amer. med. Ass. **189,** 311 (1964).

HOOPER, J. H., WELCH, V. C., POINT, P., SHACKELFORD, R. T.: Abnormal lactation associated with tranquilizing drug therapy. J. Amer. med. Ass. **178,** 506 (1961).

HORNYKIEWICZ, O.: Dopamine (3-hydroxytyramine) and brain function. Pharmacol. Rev. **18,** 925 (1966).

HOTOVY, R., KAPFF-WALTER, J.: Die pharmakologischen Eigenschaften des Perphenazinsulfoxyds. Arzneimittel-Forsch. **10,** 638 (1960).

HUANG, W. Y., PEARLMAN, J. W. H.: The corpus luteum and steroid hormone formation. Studies on luteinized rat ovarian tissue *in vitro*. J. biol. Chem. **237**, 1060 (1962).

HUNT, V. C., BUDD, J. W.: Gynecomastia associated with interstitial cell tumor of the testicle. J. Urol. (Baltimore) **42**, 1242 (1939).

IMURA, H., SPARKS, L. L., GRODSKY, G. M., FORSHAM, P. H.: Immunologic studies of adrenocorticotropic hormone (ACTH): Dissociation of biologic and immunologic activities. J. clin. Endocr. **25**, 1361 (1965).

JACOBS, E. C.: Gynecomastia following severe starvation. Ann. intern. Med. **28**, 792 (1948).

JACOBSOHN, D.: Mammary gland growth in relation to hormones with metabolic actions. Proc. roy. Soc. B. **149**, 325 (1958).

— Thyroxin and the reaction of the mammary glands to ovarian steroids in hypophysectomized rats. J. Physiol. (Lond.) **148**, 10 P (1959).

— Hormonal regulation of mammary gland growth. In: Milk, the mammary gland and its secretion. Vol. I, p. 127. Eds.: S. K. KON and A. T. COWIE. New York-London: Academic Press 1961.

— Modification by estrogens of the reaction of the rat's mammary gland to androgens. Acta endocr. (Kbh.) **41**, 88 (1962 a).

— Interactions of estrogens and androgens on the mammary gland and growth of other tissues in hypophysectomized rats treated with insulin, cortisone and thyroxine. Acta endocr. (Kbh.) **41**, 287 (1962 b).

JANSSEN, P. A. J.: Haloperidol and related butyrophenones. In: Psychopharmacological agents. Vol. II, p. 199. Ed.: M. GORDON. New York-London: Academic Press 1967 a.

— The pharmacology of haloperidol. Int. J. Neuropsychiat. **3**, Suppl. 1, 10 (1967 b).

— NIEMECEERS, C. J. E., SCHELLEKENS, K. H. L.: Is it possible to predict the clinical effects of neuroleptic drugs (major tranquilizers) from animal data? Arzneimittel-Forsch. **15**, 1196 (1965).

— VAN DER WESTERINGH, C., JAGENEAU, A. H. M., DEMOEN, P. J. A., HERMANS, B. K. F., VAN DAELE, G. H. P., SCHELLEKENS, K. H. L., VAN DER EYCKEN, C. A. M., NIEMEGEERS, C. J. E.: Chemistry and pharmacology of CNS depressants related to 4-(4-hydroxy-4-phenylpiperidino) butyrophenone. Part I Synthesis and screening data in mice. J. med. pharm. Chem. **1**, 281 (1959).

JOHNSON, R. M., MEITES, J.: Effects of cortisone, hydrocortisone, and ACTH on mammary growth and pituitary prolactin content of rats. Proc. Soc. exp. Biol. (N. Y.) **89**, 455 (1955).

— — Effects of suckling, pitocin and cervical stimulation on release of prolactin in the parturient rat. Amer. J. med. Sci. **16**, 1045 (1957).

— — Effects of cortisone acetate on milk production and mammary involution in parturient rats. Endocrinology **63**, 290 (1958).

JONGH, S. E. DE: Corpus luteum und Laktation. Acta brev. neerl. Physiol. **2**, 199 (1932).

JOSIMOVICH, J. B.: Maintenance of pseudopregnancy in the rat by synergism between human placental lactogen and chorionic gonadotrophin. Endocrinology **83**, 530 (1968).

— MACLAREN, J. A.: Presence in the human placenta and term serum of a highly lactogenic substance immunologically related to pituitary growth hormone. Endocrinology **71**, 209 (1962).

KANEMATSU, S., HILLIARD, J., SAWYER, C. H.: Effect of hypothalamic lesions on pituitary prolactin content in the rabbit. Endocrinology **73**, 345 (1963 a).

— — — Effects of reserpine on pituitary prolactin content and its hypothalamic site of action in the rabbit. Acta endocr. (Kbh.) **44**, 467 (1963 b).

— SAWYER, C. H.: Effects of intrahypothalamic implants of reserpine on lactation and pituitary prolactin content in the rabbit. Proc. Soc. exp. Biol. (N. Y.) **113**, 967 (1963 a).

— — Effects of intrahypothalamic and intrahypophysical estrogen implants on pituitary prolactin and lactation in the rabbit. Endocrinology **72**, 243 (1963 b).

KAPLAN, S. L., GRUMBACH, M. M.: Immunoassay of human chorionic "Growth Hormone-Prolactin" in serum and urine. Science **147**, 751 (1965 a).

— — Serum chorionic "Growth Hormone-Prolactin" and serum pituitary growth hormone in mother and fetus at term. J. clin. Endocr. **25**, 1370 (1965 b).

KATZ, S., MOLITCH, M., MCCANN, S. M.: Feedback of hypothalamic growth hormone (GH) implants upon the anterior pituitary (AP). Program of the 49th Meeting of the Endocrine Society, p. 23, 1967.

KHAN, M. Y., BERNSTORE, E. C.: Effect of chlorpromazine and reserpine upon pituitary function. Exp. Med. Surg. **22**, 363 (1964).

KHAZAN, N., BEN-DAVID, M., MISHKINSKY, J., KHAZEN, K., SULMAN, F. G.: Dissociation between mammotropic and sedative effects of non-hormonal hypothalamic tranquilizers. Arch. int. Pharmacodyn. **164**, 258 (1966).

— — SULMAN, F. G.: ADH-like effect of tranquilizers in amphibians. Proc. Soc. exp. Biol. (N. Y.) **112**, 490 (1963).

— PRIMO, C., DANON, A., ASSAEL, M., SULMAN, F. G., WINNIK, H. Z.: The mammotropic effect of tranquilizing drugs. Arch. int. Pharmacodyn. **136**, 291 (1962).

— SULMAN, F. G.: Melanophore-dispersing activity of reserpine in Rana frogs. Proc. Soc. exp. Biol. (N. Y.) **107**, 282 (1961).

— — WINNIK, H. Z.: Effect of reserpine on pituitary-gonadal axis. Proc. Soc. exp. Biol. (N. Y.) **105**, 201 (1960).

— — — Activity of pituitary-adrenal cortex axis during acute and chronic reserpine treatment. Proc. Soc. exp. Biol. (N. Y.) **106**, 579 (1961).

KHAZEN, K., MISHKINSKY, J., BEN-DAVID, M., SULMAN, F. G.: Lactogenic effect of phenothiazine-like drugs. Arch. int. Pharmacodyn. **174**, 428 (1968 a).

KHAZEN, K., MISHKINSKY, J., BEN-DAVID, M., SULMAN, F. G.: Structure-activity-relationship of mammotropic phenothiazine derivatives. Arch. int. Pharmacodyn. **171**, 251 (1968 b).

KILPATRICK, R., ARMSTRONG, D. T., GREEP, R. O.: Maintenance of the corpus luteum by gonadotrophins in the hypophysectomized rabbit. Endocrinology **74**, 453 (1964).

KINROSS-WRIGHT, J., CHARALAMPOUS, K. D.: A controlled study of a very long-acting phenothiazine preparation. Int. J. Neuropsychiat. **1**, 66 (1965).

KITAY, J. I., ALTSCHULE, M. D.: The pineal gland; a review of the physiologic literature. Cambridge (Mass.): Harvard University Press 1954.

KLATSKIN, G., SALTER, W. T., HUMM, F. D.: Gynecomastia due to malnutrition. 1. Clinical studies. Amer. J. med. Sci. **213**, 19 (1947).

KLEIN, J. J., SEGAL, R. L., PICHEL-WARNER, R. R.: Galactorrhea due to imipramine; Report of a case. New Engl. J. Med. **271**, 510 (1964).

KLINEFELTER, H. F., JR., REIFENSTEIN, E. C., JR., ALBRIGHT, F.: Syndrome characterized by gynecomastia, aspermatogenesis without A-leydigism, and increased excretion of follicle-stimulating hormone. J. clin. Endocr. **2**, 615 (1942).

KOANG, N. K., HU, T. C., TCHEN, C. L., CHU, T. H.: Endocrine function during treatment of pulmonary tuberculosis with INH. China med. J. **75**, 100 (1957).

KON, S. K., COWIE, A. T. (Eds.): Milk; the mammary gland and its secretion. I—II. New York-London: Academic Press 1961.

KOVACIC, N.: Luteotrophic activity of human growth hormone (Raben). Nature (Lond.) **195**, 1210 (1962 a).

— Prolongation of diestrus in the mouse as a quantitative assay of luteotrophic activity of prolactin. J. Endocr. **24**, 227 (1962 b).

— The deciduoma assay: a method for measuring prolactin. J. Endocr. **28**, 45 (1963).

— Mouse hyperaemic corpora lutea assay of prolactin. J. Reprod. Fertil. **15**, 259 (1968).

KRAGT, C. L., MEITES, J.: Stimulation of pigeon pituitary prolactin release by pigeon hypothalamic extract *in vitro*. Endocrinology **76**, 1169 (1965).

— — Dose-response relationships between hypothalamic PIF and prolactin release by rat pituitary tissue *in vitro*. Endocrinology **80**, 1170 (1967).

KRONFELD, D. S., MAYER, G. P., ROBERTSON, J. McD., RAGGI, F.: Depression of milk secretion during insulin administration. J. Dairy Sci. **46**, 559 (1963).

KUMARESAN, P., ANDERSON, R. R., TURNER, C. W.: Effect of corticosterone upon mammary gland growth of pregnant rats. Endocrinology **81**, 658 (1967).

KUPPERMAN, H. S.: Human Endocrinology. Vol. II, p. 453. Philadelphia: Davis 1963 a.

— Human Endocrinology. Vol. II, p. 604 and 644. Philadelphia: Davis 1963 b.

KUPPERMAN, H. S., FRIED, P., HAIR, L. Q.: The control of menorrhagia by prolactin. Amer. J. Obstet. Gynec. **48,** 228 (1944).

KURAMITSU, C., LOEB, L.: The effect of suckling and castration on the lactating mammary gland in rat and guinea pig. Amer. J. Physiol. **56,** 40 (1921).

KURCZ, M.: Studies on the neuro-humoral regulation of lactation. In: Symp. on Reproduction (Aug.—Sep. 1967, Pecs, Hungary), p. 175. Budapest: Akademai Kiado 1967.

KUROSHIMA, A., ARIMURA, A., BOWERS, C. Y., SCHALLY, A. V.: Inhibition by pig hypothalamic extracts of depletion of pituitary prolactin rats following cervical stimulation. Endocrinology **78,** 216 (1966).

KWA, H. G.: Personal communication (1968).

— VERHOFSTAD, F.: Radioimmunoassay of rat prolactin. Biochim. biophys. Acta (Amst.) **133,** 186 (1967).

LAHR, E. L., RIDDLE, O.: Temporary suppression of estrous cycles in the rat by prolactin. Proc. Soc. exp. Biol. (N. Y.) **34,** 880 (1936).

LARON, Z.: The immunological properties and biological activities of various growth hormones. Proc. Sec. Int. Congress Endocrin., p. 1195. Amsterdam: Excpt. Med. Publ. 1965.

— ASSA, S.: Immunological investigations of human prolactin preparation. In: Excerpta medica, international congress, Series no. 83, proceeding of the second international congress of endocrinology, London, August 1964, p. 1259. Amsterdam: Excerpta medica foundation 1964.

— — APOSTOLAKIS, M.: Immunological studies of a human prolactin preparation. Acta endocr. (Kbh.) Suppl. **100,** 164 (1965).

LASS, P. M., SMELSER, J., KURZROK, R.: Studies relating to time of human ovulation III. During lactation. Endocrinology **23,** 39 (1938).

LAUNCHBURY, A. P.: "Nuclear-modified" phenothiazines. J. Hosp. Pharm. **21,** 267 (1964).

LEATHEM, J. H.: Extragonadal factors in reproduction: In: Recent progress in the endocrinology of reproduction, p. 179. Ed.: C. W. LLOYD. New York: Academic Press 1959.

LEECH, F. B., BAILEY, G. L.: The effect on the health of lactating cows of treatment with galactopoietic doses of thyroxine or iodinated casein. J. agric. Sci. **43,** 236 (1953).

LEGORI, M., MOTTA, M., ZANISL, M., MARTINI, L.: "Short feedback loops" in ACTH control. Acta endocr. (Kbh.) Suppl. **100,** 155 (1965).

LEONARD, S. L., REECE, R. P.: The relation of the thyroid to mammary gland growth in the rat. Endocrinology **28,** 65 (1941).

LEVY, R. D., SAMPLINER, J.: Prolactin, immunologic evidence of species specificity. Proc. Soc. exp. Biol. (N. Y.) **109,** 672 (1962).

LEWIN, M. L.: Gynecomastia: The hypertrophy of the male breast. J. clin. Endocr. **1,** 511 (1941).

LEWINN, E. B.: Gynecomastia during digitalis therapy; report of eight additional cases with liver-function studies. New Engl. J. Med. **248,** 316 (1953).

LI, C. H.: Biochemistry of prolactin. In: Milk, the mammary gland and its secretion. Vol. I, p. 205. Eds.: S. K. KON and A. T. COWIE. New York-London: Academic Press 1961.
— In: Discussion on prolactin (panel). Ciba Foundation Colloq. Endocr. **14,** 360 (1962).
LIGGINS, G. C., IBBERTSON, H. K.: A successful quintuplet pregnancy following treatment with human pituitary gonadotrophin. Lancet **1966 I,** 114.
LISSEN, A. W., PARKES, M. W.: The relation between sedation and body temperature in the mouse. Brit. J. Pharmacol. 12, 245 (1957).
LONG, J. A., EVANS, H. M.: Estrous cycle in the rat and its associated phenomena. Mem. Univ. Calif. **6,** 1 (1922).
LYONS, W. R.: Lobulo-alveolar mammary growth in the rat. Colloq. int. center nat. rech. sci. **32,** 29 (1951).
— Hormonal synergism in mammary growth. Proc. roy. Soc. B. **149,** 303 (1958).
— LI, C. H., AHMAD, N., RICE-WRAY, E.: Mammotrophic effects of human hypophysial growth hormone preparations in animals and man. In: Excerpta medica, international congress series no. 158, Growth Hormone, proceedings of the first international symposium, Milan, Italy, September, 1967, p. 349. Amsterdam: Excerpta medica foundation 1968.
— — COLE, R. D., JOHNSON, R. E.: Some of the hormones required by the mammary gland in its development and function. J. clin. Endocr. **13,** 836 (1953).
— DIXON, J. S.: The physiology and chemistry of the mammotrophic hormone. In: The Pituitary Gland. London: Butterworths 1966.
— — JOHNSON, R. G.: The enhancing effect of somatotropin on the mammary growth induced in rats with estrin, progestin and mammotropin. J. clin. Endocr. **12,** 937 (1952).
— — — The hormonal control of mammary growth and lactation. Recent Progr. Hormone Res. **14,** 219 (1958).
— — — Biologic activities of human hypophysial mammotrophin. Program of the Forty-third Meeting of the Endocrine Society. New York 1961, p. 4.
— PAGE, E.: Detection of mammotropin in the urine of lactating women. Proc. Soc. exp. Biol. (N. Y.) **32,** 1049 (1935).
MAANEN, J. N. VAN, SMELIK, P. G.: Induction of pseudopregnancy in rats following local depletion of monoamines in the median eminence of the hypothalamus. Neuroendocrinology **3,** 177 (1968).
MAAS, J. M.: Amenorrhea-galactorrhea syndrome before, during, and after pregnancy. Fertil. and Steril. **18,** 857 (1967).
MCCALISTER, A., WELBOURN, R. B.: Stimulation of mammary cancer by prolactin and the clinical response to hypophysectomy. Brit. med. J. **1962 I,** 1669.
MCCANN, S. M., MACK, R., GALE, C.: The possible role of oxytocin in stimulating the release of prolactin. Endocrinology **64,** 870 (1959).

MacDonald, D., O'Driscoll, K.: Suppression of lactation, a double-blind trial. Lancet **1965 II,** 623.

McGarry, E. E., Beck, J. C.: The hyperglycaemic and hypoglycaemic effects of human growth hormone. 4ème Congrès de la Fédération internationale du Diabète, Geneva 1961. Comptes rendus. Vol. I. Genève: Editions médecine et hygiène 1961, p. 84.

— Rubinstein, D., Beck, J. C.: Growth hormones and prolactins; biochemical immunological and physiological similarities and differences. Ann. N. Y. Acad. Sci. **148,** 559 (1968).

McQueen-Williams, M.: Decreased mammotropin in pituitaries of thyroidectomized (maternalized) male rats. Proc. Soc. exp. Biol. (N. Y.) **33,** 406 (1935).

Maickel, R. P., Stern, D. N., Brodie, B. B.: In: Biochemical and neurophysiological correlation centrally acting drugs; Proceedings 2nd international pharmacological meeting, Prague, 1963. Vol. 2. Eds.: P. Paoleth and E. Trabucchi. London: Pergamon 1963, p. 225.

Malven, P. V., Hansel, W., Sawyer, C. H.: A mechanism antagonizing the luteotrophic action of exogenous prolactin in rats. J. Reprod. Fertil. **13,** 205 (1967).

— Sawyer, C. H.: A luteolytic action of prolactin in hypophysectomized rats. Endocrinology **79,** 268 (1966).

Mandl, A. M.: Corpora lutea in senile virgin laboratory rats. J. Endocr. **18,** 438 (1959).

Mann, N. M.: Gynecomastia during therapy with spironolactone. J. Amer. med. Ass. **184,** 778 (1963).

Margolis, I. B., Gross, C. G.: Gynecomastia during phenothiazine therapy. J. Amer. med. Ass. **199,** 942 (1967).

Marshall, W. K., Leiberman, D. M.: A rare complication of chlorpromazine treatment. Lancet **1956,** 270, 162.

Martini, L., Ganong, W. F.: Neuroendocrinology. Vol. I. New York-London: Academic Press 1966.

Mason, N. R., Marsh, J. M., Savard, K.: An action of gonadotropin *in vitro.* J. biol. Chem. **237,** 1801 (1962).

Matthies, D. L.: Studies of the luteotropic and mammotropic factor found in trophoblast and maternal peripheral blood of the rat at midpregnancy. Anat. Rec. **159,** 55 (1967).

Mayer, D. T.: Rat's milk and the stomach contents of suckling rats. J. Nutr. **10,** 343 (1935).

Mayer, G.: Le prolactine, facteur lutéotrophique. Arch. Sci. physiol. **5,** 247 (1951).

Mazer, C., Goldstein, L.: Clinical endocrinology of the female. Philadelphia: Saunders 1932.

Meites, J.: Le déclenchement de la lactation au moment de la parturition. Ann. Endocr. (Paris) **17,** 519 (1956).

— Induction of lactation in rabbits with reserpine. Proc. Soc. exp. Biol. (N. Y.) **96,** 728 (1957 a).

MEITES, J.: Effects of growth hormone on lactation and body growth of parturient rats. Proc. Soc. exp. Biol. (N. Y.) **96,** 730 (1957 b).
— Mammary growth and lactation. In: Reproduction in domestic animals, Vol. I. Eds.: H. H. COLE and P. T. CUPPS. New York-London: Academic Press 1959 a, p. 539.
— Hormonal prolongation of lactation for 75 days after litter removal in parturient rats. Fed. Proc. **18,** 103 (1959 b).
— Farm animals; hormonal induction of lactation and galactopoiesis. In: Milk; the mammary gland and its secretion. Vol. I, p. 321. Eds.: S. K. KON and A. T. COWIE. New York-London: Academic Press 1961.
— In: Pharmacological control of release of hormones including antidiabetic drugs; Proceedings first international pharmacological meeting, Stockholm 1961. Vol. I, p. 151. Ed.: R. GUILLEMIN. London: Pergamon 1961 a.
— Pharmacological control of prolactin secretion and lactation. Biochem. Pharmacol. **8,** 27 (1961 b).
— Neurohumoral and pharmacological control of prolactin secretion. Scientific group on the physiology of lactation. Geneva: World Health Organization 1963.
— Maintenance of the mammary lobulo-alveolar system in rats after adreno-orchidectomy by prolactin and growth hormone. Endocrinology **76,** 1220 (1965).
— Control of mammary growth and lactation. In: Neuroendocrinology, Vol. I. Eds.: L. MARTINI and W. F. GANONG. New York-London: Academic Press 1966, p. 669.
— HOPKINS, T. E.: Mechanism of action oxytocin in retarding mammary involution: study in hypophysectomized rats. J. Endocr. **22,** 207 (1961).
— — TALWALKER, P. K.: Induction of lactation in pregnant rabbits with prolactin, cortisol acetate or both. Endocrinology **73,** 261 (1963).
— KAHN, R. H., NICOLL, C. S.: Prolactin production by rat pituitary *in vitro*. Proc. Soc. exp. Biol. (N. Y.) **108,** 440 (1961).
— KRAGT, C. L.: Effects of a pituitary homotransplant and thyroxine on body and mammary growth in immature hypophysectomized rats. Endocrinology **75,** 565 (1964).
— NICOLL, C. S.: Hormonal prolongation of lactation for 75 days after litter withdrawal in portpartum rats. Endocrinology **65,** 572 (1959).
— — Adenohypophysis: Prolactin. Ann. Rev. Physiol. **28,** 57 (1966).
— — TALWALKER, P. K.: Effects of reserpine and serotonin on milk secretion and mammary growth in the rat. Proc. Soc. exp. Biol. (N. Y.) **101,** 563 (1959).
— — — Local action of oxytocin on mammary glands of postpartum rats after litter removal. Proc. Soc. exp. Biol. (N. Y.) **103,** 118 (1960).
— — — The central nervous system and the secretion and release of prolactin. Advances in neuroendocrinology, p. 238. Ed.: A. V. NALBANDOV. Urbana (Ill.): Univ. of Illinois Press 1963.

Meites, J., Sgouris, J. T.: Can the ovarian hormones inhibit the mammary response to prolactin? Endocrinology **53**, 17 (1953).

— — Effects of altering the balance between prolactin and ovarian hormones on initiation of lactation in rabbits. Endocrinology **55**, 530 (1954).

— Shelesnyak, M. C.: Effects of prolactin on duration of pregnancy, viability of young and lactation in rats. Proc. Soc. exp. Biol. (N. Y.) **94**, 746 (1957).

— Talwalker, P. K., Nicoll, C. S.: Initiation of lactation in rats with hypothalamic or cerebral tissue. Proc. Soc. exp. Biol. (N. Y.) **103**, 298 (1960 a).

— — — Failure of oxytocin to initiate mammary secretion in rabbits or rats. Proc. Soc. exp. Biol. (N. Y.) **105**, 467 (1960 b).

— Turner, C. W.: Extraction and assay of lactogenic hormone in postpartum urine. J. clin. Endocr. **1**, 918 (1941).

— — Studies concerning the mechanism controlling the initiation of lactation at parturition. I. Can estrogen suppress the lactogenic hormone of the pituitary? Endocrinology **30**, 711 (1942 a).

— — Studies concerning the mechanism controlling the initiation of lactation at parturition. II. Why lactation is not initiated during pregnancy. Endocrinology **30**, 719 (1942 b).

— — Studies concerning the mechanism controlling the initiation of lactation at parturition. III. Can estrogen account for the precipitous increase in the lactogen content of the pituitary following parturition? Endocrinology **30**, 726 (1942 c).

— — Studies concerning mechanism controlling initiation of lactation at parturition. IV. Influence of suckling on lactogen content of pituitary of postpartum rabbits. Endocrinology **31**, 340 (1942 d).

— — Effect of thiouracil and estrogen on lactogenic hormone and weight of pituitaries of rats. Proc. Soc. exp. Biol. (N. Y.) **64**, 488 (1947).

— — Studies concerning the induction and maintenance of lactation. I. The mechanism controlling the initiation of lactation at parturition. Res. Bull. Mi. agric. Exp. Stn. No. **415** (1948).

— — Lactogenic hormone. In: Hormone assay, p. 237. Ed.: C. W. Emmens. New York: Academic Press 1950.

Mena, F., Beyer, C.: Induction of milk secretion in the rabbit by lesions in the temporal lobe. Endocrinology **83**, 618 (1968).

Mercier-Parot, L.: Troubles du développement post-natal du rat après administration de cortisone à la mère gestante ou allaitante. C. R. Acad. Sci. (Paris) **240**, 2259 (1955).

Mess, B., Martini, L.: The central nervous system and the secretion of anterior pituitary trophic hormones. Recent Advances in Endocrinology. VIIIth edition, p. 1. Ed.: V. H. T. James. London: Churchill 1968.

Meyler, L. (Ed.): Side effects of drugs; adverse reactions as reported in the medical literature of the world 1963—1965. Vol. V. Amsterdam: Excerpta medica foundation 1966.

MINAGUCHI, H., CLEMENS, J. A., MEITES, J.: Changes in pituitary prolactin levels in rats from weaning to adulthood. Endocrinology **82,** 555 (1968).

— MEITES, J.: Effects of norethynodrel-mestranol combination (Enovid) on hypothalamic and pituitary hormones in rats. Endocrinology **81,** 826 (1967).

MISHKINSKY, J., DIKSTEIN, S., BEN-DAVID, M., AZEROUAL, J., SULMAN, F. G.: A sensitive *in vitro* method for prolactin determination. Proc. Soc. exp. Biol. (N. Y.) **125,** 360 (1967).

— KHAZEN, K., GIVANT, Y., SULMAN, F. G.: Mammotropic and neuroleptic effects of butyrophenones in the rat. Arch. int. Pharmacodyn. (in press) (1969).

— — SULMAN, F. G.: Prolactin-releasing activity of the hypothalamus in post-partum rats. Endocrinology **82,** 611 (1968).

— — — Mammotropic effect of fluphenazine enanthate in the rat. Neuroendocrinology **4,** 321 (1969).

— LAJTOS, Z. K., SULMAN, F. G.: Initiation of lactation by hypothalamic implantation of perphenazine. Endocrinology **78,** 919 (1966).

— NIR, I., LAJTOS, Z. K., SULMAN, F. G.: Mammary development in pinealectomized rat. J. Endocr. **36,** 215 (1966).

— — SULMAN, F. G.: Internal feedback of prolactin in the rat. Neuroendocrinology **5,** 48 (1969).

— ZANBELMAN, L., KHAZEN, K., SULMAN, F. G.: Mammotropic and prolactin-like effects of placenta and amniotic fluid of rat and human origin. J. Endocr. **46,** 15 (1970).

MITTLER, J. C., MEITES, J.: Effects of epinephrine and acetylcholine on hypothalamic content of prolactin inhibiting factor. Proc. Soc. exp. Biol. (N. Y.) **124,** 310 (1967).

MIZUNO, H., TALWALKER, P. K., MEITES, J.: Influence of hormones on tumor growth and plasma prolactin levels in rats bearing pituitary "mammotropic" tumor. Cancer Res. **24,** 1433 (1964).

MOLINA, C., ABERKANE, B., CONQUY-DOUARD, T. A. M.: Les gynécomasties chez les tuberculeux pulmonaires; à propos de cinq observations. Maroc. méd. **36,** 635 (1957).

MOON, R. C.: Mammary growth in rats treated with somatotropin during pregnancy and/or lactation. Proc. Soc. exp. Biol. (N. Y.) **118,** 181 (1965).

— Personal communication (1966).

MOORE, W. W., NALBANDOV, A. V.: Maintenance of corpora lutea in sheep with lactogenic hormone. J. Endocr. **13,** 18 (1955).

MUNFORD, R. E.: The effect of cortisol acetate on estrone-induced mammary gland growth in immature ovariectomized albino mice. J. Endocr. **16,** 72 (1957).

MURRAY, M. R.: Nervous tissues *in vitro*. Cells and tissues in culture. Vol. II, p. 373. Ed.: E. N. WILLMER. New York-London: Academic Press 1965.

NAGATA, G.: Influences of chlorpromazine on the pituitary function especially corticotropin and gonadotrophin secretion. Rep. Shionogi Res. Lab. **7**, 71 (1957).

NANDI, S.: Endocrine control of mammary gland development and function in the C3H/Hc CRGL mouse. J. nat. Cancer Inst. **21**, 1039 (1958 a).

— Role of somatotropin in mammogenesis and lactogenesis in C3H/Hc CRGL mice. Science **128**, 772 (1958 b).

— Hormonal control of mammogenesis and lactogenesis in the C_3H/He CRGL mouse. Univ. Calif. Publ. Zool. **65**, 4 (1959).

— BERN, H. A.: The hormones responsible for lactogenesis in BALB/c CRGL mice. Gen. comp. Endocr. **1**, 195 (1961).

NELSON, W. O.: Endocrine control of the mammary gland. Physiol. Rev. **16**, 488 (1936).

— GAUNT, R., SCHWEIZER, M.: Effects of adrenal cortical compounds on lactation. Endocrinology **33**, 325 (1943).

NICOLL, C. S.: Neural regulation of adenohypophysical prolactin secretion in tetrapods: indications from *in vitro* studies. J. exp. Zool. **158**, 203 (1965).

— Bio-assay of prolactin. Analysis of the pigeon crop-sac response to local prolactin injection by an objective and quantitative method. Endocrinology **80**, 641 (1967).

— Electrophoretic comparison of the tetrapod prolactins. Excerpta medica, international congress series no. 157, third international congress of endocrinology, Mexiko, D. F., June 30—July 5, 1968, p. 77. Amsterdam: Excerpta medica foundation 1968.

— Bio-assay of prolactin. Acta endocr. (Kbh.) **60**, 91 (1969).

— Assay of hypothalamic factors which regulate prolactin secretion. In: Hypophysiotropic hormones of the hypothalamus: assay and chemistry, p. 115. Ed.: J. MEITES. Baltimore: Williams & Wilkins Publ. 1970.

— BERN, H. A.: "Prolactin" and the pituitary glands of fishes. Gen. comp. Endocr. **4**, 457 (1964).

— MEITES, J.: Prolongation of lactation in the rat by litter replacement. Proc. Soc. exp. Biol. (N. Y.) **101**, 81 (1959).

— — Estrogen stimulation of prolactin production by rat adenohypophysis *in vitro*. Endocrinology **70**, 272 (1962 a).

— — Failure of neurohypophysical hormones to influence prolactin secretion *in vitro*. Endocrinology **70**, 927 (1962 b).

— — Prolactin secretion *in vitro:* comparative aspects. Nature (Lond.) **195**, 606 (1962 c).

— — Prolactin secretion *in vitro:* effects of thyroid hormones and insulin. Endocrinology **72**, 544 (1963).

— — Prolactin secretion *in vitro:* effects of gonadal and adrenal cortical steroids. Proc. Soc. exp. Biol. (N. Y.) **117**, 579 (1964).

— TALWALKER, P. K., MEITES, J.: Initiation of lactation in rats by nonspecific stresses. Amer. J. Physiol. **198**, 1103 (1960).

NIKITOVITCH-WINER, M. B., EVERETT, J. W.: Comparative study of luteotropin secretion by hypophysial autotransplants in the rat. Effects of site and stages of the estrus cycle. Endocrinology **62**, 522 (1958).

NIR, I., MISHKINSKY, J., ESHCHAR, N., SULMAN, F. G.: Effect of pinealectomy on milk yield in rats. J. Endocr. **42**, 161 (1968).

NISWENDER, G. D., CHEN, C. L., MIDGLEY, A. R., JR., MEITES, J., ELLIS, S.: Radioimmunoassay for rat prolactin. Proc. Soc. exp. Biol. (N. Y.) **130**, 793 (1969).

PALLADINO, A., BUTCHER, R. L., FUGO, N. W.: An additional effect of yohimbine-HCl. Proc. Soc. exp. Biol. (N. Y.) **127**, 408 (1968).

PALOYAN, E., KOLAR, J. B., ERNST, K., PALOYAN, D., HARPER, P. V.: The effect of glucagon on serum calcium. Fed. Proc. **24**, 323 (1965).

— PALOYAN, D., HARPER, P. V.: Glucagon-induced hypocalcemia. Metabolism **16**, 35 (1967).

PARKES, C. M., BROWN, G. W., MONCK, E. M.: The general practitioner and the schizophrenic patient. Brit. med. J. **1962 I**, 972.

PASTEELS, J. L.: Premiers résultats de culture combinée *in vitro* d'hypophysis et d'hypothalamus, dans le but d'en apprécier la sécrétion de prolactin. C. R. Acad. Sci. (Paris) **253**, 3074 (1961 a).

— Sécrétion de prolactine par l'hypophyse en culture de tissus. C. R. Acad. Sci. (Paris) **253**, 2140 (1961 b).

— Administration d'extraits hypothalamiques à l'hypophyse de rat *in vitro*, dans le but d'en contrôler la sécrétion de prolactine. C. R. Acad. Sci. (Paris) **254**, 2664 (1962).

— Recherches morphologiques et expérimentales sur la sécrétion de prolactine. Arch. Biol. (Liège) **74**, 440 (1963).

— Hormone de croissance et prolactine dans l'hypophyse humaine. Ann. Endocr. (Paris) **28**, 117 (1967).

— BRAUMAN, H., BRAUMAN, J.: Étude comparé de la sécretion d'hormone somatotrope par l'hypophyse humaine *in vitro*, et son activité lactogénique. C. R. Acad. Sc. (Paris) **256**, 2031 (1963).

PAVEL, S., PETRESCU, S.: Inhibition of gonadotrophin by a highly purified pineal peptide and by synthetic arginine vasotocin. Nature (Lond.) **212**, 1054 (1966).

PEAKE, G. T., MCKEEL, D. W., JARETT, L., DAUGHADAY, W. H.: Ultrastructural histologic and hormonal characterization of a prolactin-rich human pituitary tumor. Endocrinology **29**, 1383 (1969).

PETERSEN, W. E.: Lactation. Physiol. Rev. **24**, 340 (1944).

PETERSON, P. V., NIELSON, I. M.: Thioxanthene derivatives. Psychopharmacological agents. Vol. I, p. 301. Ed.: M. GORDON. New York-London: Academic Press 1964.

PETTINGER, W. A., HORWITZ, D., SJOERDSMA, A.: Lactation due to methyldopa. Brit. med. J. **1963 I**, 1460.

PLATT, S. S., SCHULZ, R. Z., KUNSTADTER, R. H.: Hypertrophy of male breast associated with recovery from starvation. Bull. U.S. Army med. Dep. **7**, 403 (1946).

PLATT, R., SEARS, H. T. N.: Reserpine in severe hypertension. Lancet **1956 I,** 401.

PLUMER, A. J., EARL, A., SCHNEIDER, J. A., TRAPOLD, J., BARRETT, W.: Pharmacology of *Rauwolfia* alkaloids, including reserpine. Ann. N.Y. Acad. Sci. **59,** 8 (1954).

POLISHUK, W. Z., KULCSAR, S.: Effects of chlorpromazine on pituitary function. J. clin. Endocr. **16,** 292 (1956).

POPOVICH, S.: Prolactin releasing factor. Dokl. Akad. Nauk SSSR, Dtd. Biol. **121,** 186 (1958).

PROP, F. J. A.: Effects of hormones on mouse mammary glands *in vitro.* Analysis of the factors that cause lobulo-alveolar development. Path. et Biol. **9,** 640 (1961 a).

— Hormones and mammary tissue *in vitro.* Acta physiol. pharmacol. neerl. **10,** 305 (1961 b).

— Sensitivity to prolactin of mouse mammary glands *in vitro.* Exp. Cell Res. **24,** 629 (1961 c).

PURSHOTTAM, N.: Effects of tranquilizers on induced ovulation in mice. Amer. J. Obstet. Gynec. **83,** 1405 (1962).

QUINN, D. L., EVERETT, J. W.: Delayed pseudopregnancy induced by selective hypothalamic stimulation. Endocrinology **80,** 155 (1967).

QUINLAN, J. J.: Lactation following thoracotomy. Canad. J. Surg. **11,** 60 (1968).

RACADOT, J.: Neurosécrétion et activité thyroidienne chez la chatte au cours de la gestation et de l'allaitement. Ann. Endocr. (Paris) **18,** 628 (1957).

RAJKOVITZ, K.: Localization of pituitary function in the diencephalon. In: Hypothalamic control of the anterior pituitary. Eds.: J. STENTAGOTHAI, B. FLERKO, B. MESS and B. HALASZ. Budapest: Akademiai Kaido 1962.

RALSTON, N. P., COWSERT, W. C., RAGSDALE, A. C., HERMAN, H. A., TURNER, C. W.: The yield and composition of the milk of dairy cows and goats as influenced by thyroxine. Res. Bull. Mo. agric. Exp. Stn. No. **317,** 1940.

RAMIREZ, V. D., MCCANN, S. M.: Induction of prolactin secretion by implants of estrogen into the hypothalamo-hypophysial region of female rats. Endocrinology **75,** 206 (1964).

RATNER, A., MEITES, J.: Depletion of prolactin-inhibiting activity of rat hypothalamus by estradiol or suckling stimulus. Endocrinology **75,** 377 (1964).

— TALWALKER, P. K., MEITES, J.: Effect of estrogen administration *in vivo* on prolactin release by rat pituitary *in vitro.* Proc. Soc. exp. Biol. (N.Y.) **112,** 12 (1962).

— — — Effect of reserpine on prolactin-inhibiting activity of rat hypothalamus. Endocrinology **77,** 315 (1965).

RAY, E. W., AVERILL, S. C., LYONS, W. R., JOHNSON, R. E.: Rat placental hormonal activities corresponding to those of pituitary mammotropin. Endocrinology **56,** 359 (1955).

REECE, R. P.: Lactogen content of female guinea pig pituitary. Proc. Soc. exp. Biol. (N. Y.) **42,** 54 (1939).

— TURNER, C. W.: The lactogenic and thyrotropic hormone content of the anterior lobe of the pituitary gland. Mo. agric. Exp. Sta. Res. Bull. **266** (1937).

REICHLIN, S.: Functions of the median-eminence gland. New Engl. J. Med. **275,** 600 (1966).

REINEKE, E. P.: Unpublished observations (1942) quoted by MEITES, J.: Thyroactive substances. In: Milk, the mammary gland and its secretion. Vol. I, p. 353. Eds.: S. K. KON ond A. T. COWIE. New York-London: Academic Press 1961.

REISFELD, R. A., TONG, G. L., RICKES, E. L., BRINK, N. G., STEELMAN, S. L.: Purification and characterization of sheep prolactin. J. Amer. chem. Soc. **83,** 3717 (1961).

RIDDLE, O., BATES, R. W., DYKSHORN, S. W.: The preparation, identification and assay of prolactin — a hormone of the anterior pituitary. Amer. J. Physiol. **105,** 191 (1933).

RIMOIN, D. L., HOLZMAN, B., MERIMEE, J., RABINOWITZ, D., BARNES, C., TYSON, J. E. A., MCKUSICK, V. A.: Lactation in the absence of human growth hormone. J. clin. Endocr. **28,** 1183 (1968).

ROBBOY, S. J., KAHN, R. H.: Zone electrophoresis and prolactin activity of rat adenohypophysis cultivated *in vitro*. Endocrinology **78,** 440 (1966).

ROBINSON, B.: Breast changes in the male and female with chlorpromazine or reserpine therapy. Med. J. Aust. **44,** 239 (1957).

ROSSI, G. V.: The psychotherapeutic agents. Amer. J. Pharm. **136,** 6 (1964).

ROTH, J., GLICK, S. M., CUATRECASAS, P., HOLLANDER, C. S.: Acromegaly and other disorders of growth hormone secretion. Amer. intern. Med. **66,** 760 (1967).

ROTHCHILD, I.: The corpus luteum-hypophysis relationships; the luteolytic effect of luteinizing hormone (LH) in the rat. Acta endocr. (Kbh.) **49,** 107 (1965).

SALKIN, D., DAVIS, E. W.: Lactation following thoracoplasty and pneumonectomy. J. thorac. Surg. **18,** 580 (1949).

SAR, M., MEITES, J.: Changes in pituitary prolactin release and hypothalamic PIF content during the estrus cycle of rats. Proc. Soc. exp.

SAWYER, C. H.: Induction of lactation in the rabbit with reserpine. Anat. Rec. **127,** 362 (1957).

SCHALLY, A. V., KUROSHIMA, A., ISHIDA, Y., REDDING, T. W., BOWERS, C. Y.: The presence of prolactin inhibiting factor (PIF) in extracts of beef, sheep and pig hypothalami. Proc. Soc. exp. Biol. (N. Y.) **118,** 350 (1965). Biol. (N. Y.) **125,** 1018 (1967).

— — Effects of progesterone, testosterone, and cortisol on hypothalamic prolactin-inhibiting factor and pituitary prolactin content. Proc. Soc. exp. Biol. (N. Y.) **127,** 426 (1968).

SCHALLY, A. V., MEITES, J., BOWERS, C. Y., RATNER, A.: Identity of prolactin inhibiting factor (PIF) and luteinizing hormone-releasing factor (LRF). Proc. Soc. exp. Biol. (N. Y.) **117**, 252 (1964).

— STEELMAN, S. L., BOWERS, C. Y.: Effect of hypothalamic extracts on release of growth hormone *in vitro*. Proc. Soc. exp. Biol. (N. Y.) **119**, 208 (1965).

SCHAMS, D., KARG, H.: Radioimmunologische Bestimmung von Prolaktin im Blutserum vom Rind. Milchwissenschaft **24**, 263 (1969).

SCHARRER, E., SCHARRER, B.: Neuroendocrinology. New York: Columbia Univ. Press 1963.

SCHLITTLER, E., PLUMMER, A. J.: Tranquilizing drugs from Rauwolfia. In: Psychopharmacological agents. Vol. I, p. 9. Ed.: M. GORDON. New York-London: Academic Press 1964.

SEBAOUN, J., DELZANT, G., PEILLON, F.: Les tumeurs à prolactines. Bull. Soc. med. Hôp. Paris **118**, 525 (1967).

SEGALOFF, A., STEELMAN, S. L.: The human gonadotropins. Recent Progr. Hormone Res. **15**, 127 (1959).

SELYE, H.: Influence of the uterus on ovary and mammary gland. Proc. Soc. exp. Biol. (N. Y.) **31**, 488 (1934 a).

— On the nervous control of lactation. Amer. J. Physiol. **107**, 535 (1934 b).

— The effect of cortisol upon the mammary glands. Acta endocr. (Kbh.) **17**, 394 (1954).

— Der derzeitige Stand der Stressforschung. In: Psyche und Hormon. Eds.: C. A. JOËL, H. MENG, P. PARIN, H. SELYE, and F. G. SULMAN. Bern: Huber 1960, p. 125.

— COLLIP, J. B., THOMSON, D. L.: Nervous and hormonal factors in lactation. Endocrinology **18**, 237 (1934).

SHAW, J. C.: The hypophysical growth hormone, nature and actions. New York: McGraw-Hill 1955, p. 486.

— CHUNG, A. C., BUNDING, I.: The effect of pituitary growth hormone and adrenocorticotropic hormone on established lactation. Endocrinology **56**, 327 (1955).

SHELESNYAK, M. C.: Fall in uterine histamine associated with ovum implantation in pregnant rat. Proc. Soc. exp. Biol. (N. Y.) **100**, 380 (1959).

SILVER, H. K.: Juvenile hypothyroidism with precocious sexual development. J. clin. Endocr. **18**, 886 (1958).

SIMKIN, B., ARCE, R.: Prolactin activity in blood during the normal human menstrual cycle. Proc. Soc. exp. Biol. (N. Y.) **113**, 485 (1963).

— GOODART, D.: Preliminary observations on prolactin activity in human blood. J. clin. Endocr. **20**, 1095 (1960).

SLUYSER, M., LI, C. H.: Studies of pituitary lactogenic hormone. XXIII. Isolation of the monomer of ovine prolactin. Arch. Biochem. **104**, 50 (1964).

SMITH, T. C., BRAVERMAN, L. E.: The action of desoxycorticosterone acetate on the mammary gland of the immature ovariectomized rat. Endocrinology **52**, 311 (1953).

SMITH, V. R., STOTT, G. H., WALKER, C. W.: Observations on parathyroidectomized goats. J. Anim. Sci. **16,** 312 (1957).
SPELLACY, W. N., CARLSON, K. L., SCHADE, S. L.: Human growth hormone studies in patients with galactorrhea (Ahumada-del Castillo syndrome). Amer. J. Obstet. Gynec. **100,** 84 (1968).
SPIES, H. G., HILLIARD, Z., SAWYER, C. H.: Maintenance of corpora lutea and pregnancy in hypophysectomized rabbits. Endocrinology **83,** 354 (1968).
STEINER, W. G., HIMWICH, H. E.: Effects of antidepressant drugs on limbic structures of rabbit. J. nerv. ment. Dis. **137,** 277 (1963).
— — Electroencephalographic changes following administration of N-dimethylacetamide and other antitumor agents to rabbits. Int. J. Neuropharmacol. **3,** 327 (1964).
STRICKER, P., GRUETER, F.: Action du lobe antérieur de l'hypophyse sur la montée laiteuse. C.R. Soc. Biol. (Paris) **99,** 1978 (1928).
— — Recherches expérimentelles sur les fonctions du lobe antérieur de l'hypophyse; influence des extraits du lobe antérieur sur l'appareil génital de la lapine, et sur la montée laiteuse. Presse méd. **37,** 1268 (1929).
SULMAN, F. G.: Simple test for blood-ACTH. Lancet **1952 I,** 1161.
— Chromatophorotropic activity of human blood: review of 1,200 cases. J. clin. Endocr. **16,** 755 (1956 a).
— Experiments on the mechanism on the "push and pull" principle. J. Endocr. **14,** XXVII (1956 b).
— The mechanism of the "Push and Pull" principle. II. Endocrine effects of hypothalamus depressants of the phenothiazine group. Arch. int. Pharmacodyn. **118,** 298 (1959).
— TWERSKY, J.: Lack of effect of prolactin on milk production of cows. Refuah Veterinarith (J. Isr. Vet. Med. Ass.) **5,** 53, 73 (1948).
— WINNIK, H. Z.: Hormonal effects of chlorpromazine. Lancet **1956,** 270, 161.
SUSSMAN, R. M.: Sprironolactones and gynecomastia. Lancet **1963 I,** 58.
SWANSON, E. W.: The effect upon milk production and body weight of varying withdrawal periods after thyroactive supplement feeding. J. Dairy Sci. **37,** 1212 (1954).
TALEISNIK, S., ORIAS, R., OLMOS, J.: Topographic distribution of the melanocyte-stimulating hormone-releasing factor in rat hypothalamus. Proc. Soc. exp. Biol. (N. Y.) **122,** 325 (1966).
— TOMATIS, M. E.: Melanocyte-stimulating hormone-releasing and inhibiting factors in two hypothalamic extracts. Endocrinology **81,** 819 (1967).
TALWALKER, P. K., MEITES, J.: Mammary lobulo-alveolar growth induced by anterior pituitary hormones in adreno-ovariectomized and adreno-ovariectomized-hypophysectomized rats. Proc. Soc. exp. Biol. (N. Y.) **107,** 880 (1961).
— NICOLL, C. S., MEITES, J.: Induction of mammary secretion in pregnant rats and rabbits by hydrocortisone acetate. Endocrinology **69,** 802 (1961).

TALWALKER, P. K., RATNER, A., MEITES, J.: *In vitro* inhibition of pituitary prolactin synthesis and release by hypothalamic extract. Amer. J. Physiol. **205**, 213 (1963).

TEEL, H. M.: The effect of injecting anterior hypophysical fluid on the course of gestation in the rat. Amer. J. Physiol. **79**, 170 (1926).

THIÉBLOT, L.: Le rôle functionnel de l'épiphyse. Rev. Canad. Biol. **13**, 189 (1954).

THOMAS, J. W.: Hormonal relationships and applications in the production of meat, milk and eggs. Agricultural Board, Natl. Res. Council Publ. No. 266, p. 47. Washington, D.C. 1953.

— MOORE, L. A.: Thyroprotein feeding to dairy cows during successive lactations. J. Dairy Sci. **36**, 657 (1953).

TINDAL, J. S.: The lactogenic response of the rabbit to reserpine. J. Endocr. **18**, XXXVI (1959).

— KNAGGS, G. S., TURVEY, A.: Central nervous control of prolactin secretion in the rabbit: effect of local oestrogen implants in the amygdaloid complex. J. Endocr. **37**, 279 (1967).

TOPKINS, P.: The histologic appearance of the endometrium during lactation amenorrhea and its relationship to ovarian function. Amer. J. Obstet. Gynec. **45**, 48 (1943).

TÖRÖK, B.: Structure of the vascular connections of the hypothalamo-hypophysial region. Acta anat. (Basel) **59**, 84 (1964).

TRENTIN, J. J., HURST, V., TURNER, C. W.: Thiouracil and mammary growth. Proc. Soc. exp. Biol. (N. Y.) **67**, 461 (1948).

TUCKER, H. A., MEITES, J.: Induction of lactation in pregnant heifers with 9-fluoro-prednisolone acetate. J. Dairy Sci. **48**, 403 (1965).

TURKINGTON, R. W., TOPPER, Y. J.: Stimulation of casein synthesis and histological development of mammary gland by human placental lactogen *in vitro*. Endocrinology **79**, 175 (1966).

TURNER, C. W.: The mammary glands. Sex and internal secretions; a survey of recent research, 2nd ed., p. 740. Ed.: E. ALLEN. Baltimore: Williams & Wilkins 1939.

— The mammary gland. In: The anatomy of the udder of cattle and domestic animals. Columbia (Missouri): Lucas 1952.

— Regulation of lactation. A conference on radioactive isotopes in agriculture, Jan. 1956, East Lansing, Michigan, p. 403. Washington D. C.: U.S. Atomic Energy Commission 1956.

— GRIFFITH, D. R.: Possible mechanism governing release of lactogenic hormone. Excerpta medica, international congress series no. 47, XXII, Congress of International physiological sciences, Leiden, September 1962, vol. I. Lectures and symposia, part II, Symposia XI—XXII, p. 740. Amsterdam: Excerpta medica foundation 1962.

TURNER, R. A., PIERCE, J. G., DU VIGNEAUD, V.: The purification and the amino acid content of vasopressin preparations. J. biol. Chem. **191**, 21 (1951).

VELARDO, J. T.: Induction of pseudopregnancy in adult rats with Trilafon, a highly potent tranquilizer of low toxicity. Fertil. and Steril. **9,** 60 (1958).

— DAWSON, A. B., OLSEN, A. G., HISAW, F. L.: Sequence of histological changes in the uterus and vagina of the rat during prolongation of pseudopregnancy associated with the presence of deciduomata. Amer. J. Anat. **93,** 273 (1953).

VOKAER, R.: Sur l'existence d'un stade fugace a vacuoles basales (lucid zone) dans le cycle endométrial de la rate adulte. Ann. Endocr. (Paris) **11,** 652 (1950).

VOOGT, J. L., CLEMENS, J. A., MEITES, J.: Stimulation of pituitary FSH release in immature female rats by prolactin implant in median eminence. Neuroendocrinology **4,** 157 (1969).

WALKER, S. M., MATTHEWS, J. I.: Observations on the effects of prepartal and postpartal estrogen and progesterone treatment on lactation in the rat. Endocrinology **44,** 8 (1949).

WELLER, C. P.: Isolation of hypothalamic prolactin releasing factor. Bull. Res. Coun. Israel (Exp. Med.) **10E,** 237 (1963).

— MISHKINSKY, J., SULMAN, F. G.: Non-specificity of prolactin assay based on luteotrophic effect in mature mice. J. Endocr. **42,** 485 (1968).

WELSH, A. L.: The newer tranquilizing drugs. Med. Clin. N. Amer. **48,** 459 (1964).

WHITE, C.: The use of ranks in a test of significance for comparing two treatments. Biometrics **8,** 33 (1952).

WILHELMI, A. E.: Fractionation of human pituitary glands. Canad. J. Biochem. **39,** 1959 (1961).

WILKINS, L.: A feminizing adrenal tumor causing gynecomastia in a boy of five years contrasted with a virilizing tumor in a five-year-old girl. J. clin. Endocr. **8,** 111 (1948).

WILKINS, R. W.: Clinical usage of *Rauwolfia* alkaloids, including reserpine (serpasil). Ann. N. Y. Acad. Sci. **59,** 36 (1954).

WINNIK, H. Z., SULMAN, F. G.: Hormonal depression due to treatment with Chlorpromazine. Nature (Lond.) **178,** 365 (1956).

— TENNENBAUM, L.: Apparition de galactorrhée au cours du traitement de largactil. Presse méd. **63,** 1092 (1955).

WIRTH, W., GÖSSWALD, R., HÖRLEIN, U., RISSE, K. H., KREISKOTT, H.: Zur Pharmakologie acylierter Phenothiazinderivate. Arch. int. Pharmacodyn. **115,** 1 (1958).

WOLTHUIS, O. L.: An assay of prolactin based on a direct effect of this hormone on cells of the corpus luteum. Acta Endocr. (Kbh.) **42,** 364 (1963).

WOOLMAN, A. M.: Reserpine in severe hypertension. Lancet **1956 I,** 690.

WRENN, T. R., BITMAN, J., DE LAUDER, W. R., MENCH, M. L.: Influence of the placenta in mammary gland growth. J. Dairy Sci. **49,** 183 (1966).

YAMAMOTO, R., FUJISAWA, S.: Photodegradation of phenothiazine derivatives in aqueous solutions. International Congress Pharmaceutical Sciences, Muenster, Germany, September 1963.

ZAKS, M. G.: The motor apparatus of the mammary gland. Springfield (Ill.): Charles C. Thomas 1962.

ZETLER, G.: Cataleptic state and hypothermia in mice, caused by central cholinergic stimulation and antagonized by anticholinergic and antidepressant drugs. Int. J. Neuropharmacol. **7**, 325 (1968).

Subject Index

Typesetting, printing and binding: Konrad Triltsch, Graphischer Betrieb, 87 Würzburg, Germany

Monographs on Endocrinology

Already Published:

Vol. 1: OHNO, S., Sex Chromosomes and Sex-linked Genes. With 33 figures. X, 192 pages. 1967. DM 38,—; US $ 9.50.

Vol. 2: EIK-NES, K. B., and E. C. HORNING, Gas Phase Chromatography of Steroids. With 85 figures. XV, 382 pages. 1968. DM 38,—; US $ 9.50.

In Preparation:

BAULIEU, E. E., Bicêtre: Current Problems in the Metabolism of Steroid Hormones.

BERSON, S. A./YALOW, R. S., New York: Immunoassay of Peptide Hormones.

BORTH, R., Toronto: Clinical Hormone Assays.

BREUER, H./RAO, G. S., Bonn: Metabolism of Estrogens.

CALDEYRO-BARCIA, R., Montevideo: Oxytocin.

COPP, D. H., Vancouver: Calcitonin.

EDELMAN, I. S., San Francisco: Aldosterone: Gene Action and Sodium Transport.

FEDERLIN, K., Ulm: Immunopathology of Insulin. Clinical and Experimental Studies.

GREGORY, R. A., Liverpool/GROSSMAN, M. I., Los Angeles: The Gastrins.

GURPIDE, E., Minneapolis: Tracer Methods in Hormone Research.

HORTON, W. E., London: Prostaglandins.

JENSEN, E./DESOMBRE, E., Chicago: Receptors and Action of Estradiol.

LUNENFELD, B., Tel Hashomer: Gonadotropins.

MCKENZIE, J. M., Montreal: The Pathogenesis of Graves' Disease.

MÜLLER, J., Zürich: Regulation of Aldosterone Biosynthesis.

NEUMANN/STEINBECK/ELGER, Berlin: Hormones in Sexual Differentiation.

RENOLD, A. E./STAUFFACHER, W., Geneva: Pathophysiology of Diabetes.

ROBERTS, S., Los Angeles: Subcellular Mechanisms in the Regulation of Corticosteroidogenesis.

SHORT, R. V., Cambridge: The Ovary and its Hormones.

WESTPHAL, U., Louisville: Steroid-Protein Interactions.

WILLIAMS-ASHMAN, H. G., Chicago: Androgen Action.

MIX
Papier aus verantwortungsvollen Quellen
Paper from responsible sources
FSC® C105338

If you have any concerns about our products,
you can contact us on
ProductSafety@springernature.com

In case Publisher is established outside the EU,
the EU authorized representative is:
Springer Nature Customer Service Center GmbH
Europaplatz 3, 69115 Heidelberg, Germany

Printed by Libri Plureos GmbH
in Hamburg, Germany